The Man in My Head Has Lost His Mind (What Is Consciousness?)

A Thought Experiment Exploring Consciousness and Free Will, Identity and the Moral Self

(Book I in the Sentience series)

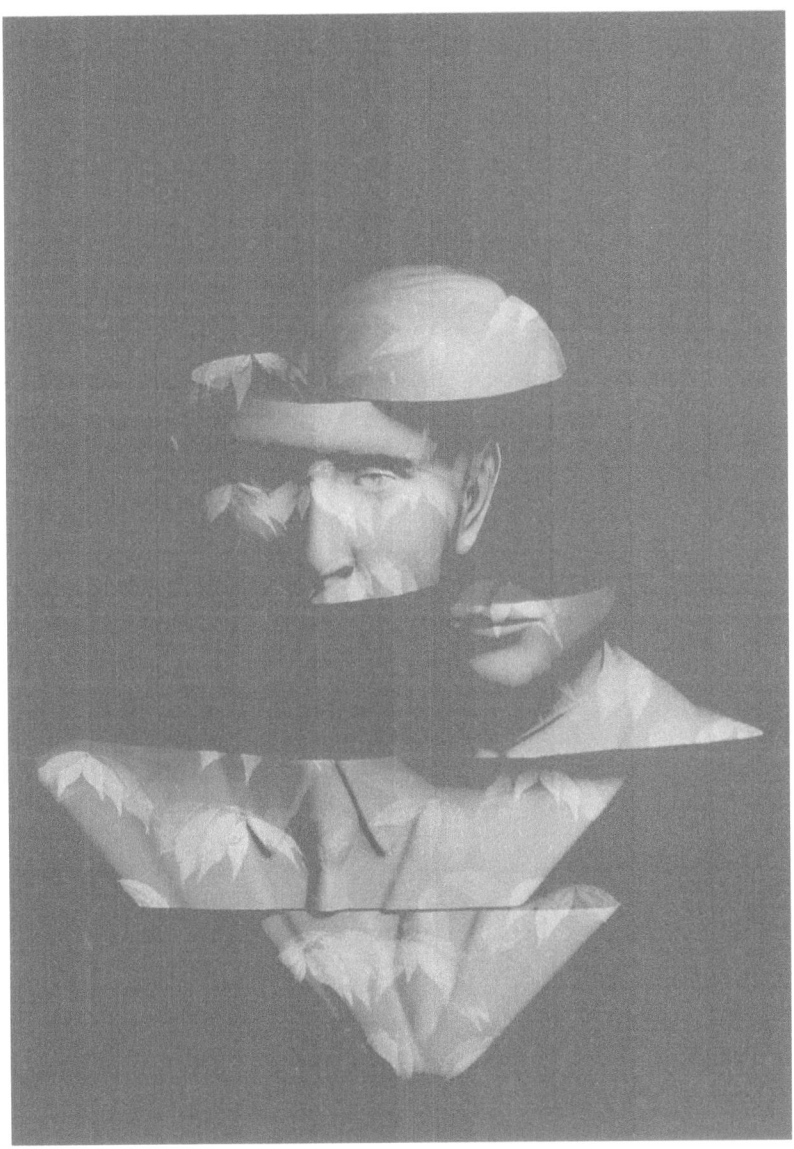

The Man in My Head Has Lost His Mind (What Is Consciousness?)

A Thought Experiment Exploring Consciousness and Free Will, Identity and the Moral Self

Carter Blakelaw

The Logic of Dreams

The Man in My Head Has Lost His Mind (What Is Consciousness?):
A Thought Experiment Exploring Consciousness, Identity and the Moral Self

First print edition. January 2023.

ISBN paperback: 9798370188626 and 9781739688783

ISBN hardback: 9798370190339 and 9781739688776 (dust jacket)

© 2022, Carter Blakelaw. All rights reserved.

No part of this publication may be reproduced, stored in a retrieval system, or transmitted, in any form or by any means, without the prior written permission of the publisher.

Published by The Logic of Dreams

Requests to publish work from this book should be sent to:

toolbox@carterblakelaw.com

While every precaution has been taken in the preparation of this book, the publisher assumes no responsibility for errors or omissions, or for damages resulting from the use of the information contained herein.

Cover art, book design and illustrations by Jack Calverley.

Photography by Antipolygon and Swapnil Dwivedi from www.unsplash.com.

10 9 8 7 6 5 4 3 2 1

www.thelogicofdreams.com

t-29-pb

The Solipsist

I called for help, but no one came,
I cried for love, that too in vain,
I sought the food to feed my mind,
But only found an unkind mess of fools.

- C.B. 2022

This text is dedicated to the memory of Jim Warren, Artist and Philosopher 1961-1990

Contents

xi	Introduction
1	1. A Poodle Ate My Homework
9	2. Life As a Comic Strip
21	3. How Do You Explain Anything?
33	4. As Time Goes By (A Kiss Is Just a Kiss)
37	5. This See, Is the Conscious Bit
45	6. Qualia, the Possible and the Particular
57	7. Evolution and Free Will
65	8. The Good, the Bad and the Choosy
69	9. Finally, Making It All Work
78	Acknowledgments
78	About the Author

Introduction

It's easy to ask a question that cannot be answered:
   "How long is a piece of string?"
   The question is well-formed and grammatical but, well, just how long *is* a piece of string?

By contrast it seems easier to answer a question like:
   "What is gravity?"
   —*Gravity is a force between two objects that have mass.*
   "What is a force?"
   —*A force is something which tends to confer motion on a mass.*
   "What is motion?"
   —*Motion is the ongoing state of changing one's location in space.*
   "What is space?"
   —*Space is a thing full of, ahem, locations...*
   "And a location?"
   —*a thing space has a lot of...*

It seems there is only so far down the rabbit hole we can go. And when it comes to consciousness the situation looks worse.
   "What is consciousness?"
   —*Consciousness is awareness.*
   "What is awareness?"
   —*Awareness is a feeling, or a collection of feelings.*
   "What is a feeling?"
   —*Something I am aware of...*

Our answers become circular. Even if we break consciousness down into sensations, thoughts and emotions, we seem never to get past *something I am aware of*. We rely on something already mentioned and not fully explained; our explanation ends up being circular. We *beg the question* we originally asked.

In this short work, which is written in the same spirit as "The Emperor's New Mind: Concerning Computers, Minds and the Laws of Physics" by Roger Penrose (OUP 2016), I suggest a way to drill down past all the question-begging and expose a satisfactory answer to the question of what consciousness is—as good a theory as any theory of gravity.

As for *How long is a piece of string?*
"----*this long*----" I promise.

Carter Blakelaw
December 2022

This book is Book I in the Sentience series. Book II in the series is **This Robot Brain Gets Life (Making AI Pseudo-Conscious)** - Design Alignment In, design Hallucination Out, and Book III in the series is **Authentic Art in the Age of AI - a de-manifesto.**

1. A Poodle Ate My Homework

Yesterday I was walking in the woods, following the path of a disused railway (long ago stripped of rails and sleepers) when a black poodle came into view. It paused in the middle of the path, stretched its neck in my direction and, with muzzle raised, sniffed the air. After a few seconds it set about trotting towards me, with confidence, not in a threatening way but, I would say, with *hope*.

Intrigued, I kept walking, and maintained eye contact. When it came to a little over a metre from me, it stopped.

It was not wagging its tail.

But nor was it growling or baring its teeth. Instead, it looked me in the eye as if it expected to find someone 'at home' (I believe also, though this may be fanciful, that it ever so slightly tilted its head).

In a friendly *can-I-help-you* voice, as one might use to address a small stray child, I said, "Hello?"

The poodle turned round and trotted off, back in the direction it came from.

I wondered briefly whether it was short-sighted and had lost its owner and for all of one minute it had considered me the lucky incumbent (I discovered at least part of the truth later). But the lasting impression I had was that when this dog looked me in the eye, it *expected* there to be something going on behind my eyes to which it could appeal.

No doubt some will dismiss my observation

automatically, call it anthropomorphism[1], and be done with it. Sure. But to you I say *you make a case against yourself*. Should I dismiss—with equal brusqueness—your presence[2], and the presence of anyone who takes a similar line, as mere affectations (good or bad) of my own mind?

However, suppose my interpretation captures something of the truth (and if you disagree strongly while insisting that *you* exist, let me invite you to consider everything in this short work as a thought experiment so that you can read further without getting riled and, possibly, even enjoy my floundering naivety).

Suppose the poodle was making an appeal to the living, breathing, thinking person that is me? What did it think it was appealing to?

Did it implicitly understand that I was, like it, conscious? I was not merely a robot capable of doing things that it could not do for itself (such as tin-opening). Did it have some sense of the mysterious me-ness[3] of me and somehow reasoned[4], or assumed, it could only access

---

[1] For our purposes, we might contemplate a range of anthropomorphic errors. They might be elaborated as strong, mixed or weak: strong when an aware observer attributes mental states to an aware observed (human to human); weak when an unaware observer infers the future behaviour of an unaware observed as if the observed had mental states (robot to robot); mixed when an unaware observer attributes mental states to an aware observed (robot to human) or when an aware observer attributes mental states to an unaware observed (human to robot). 'Mixed: aware on unaware' would apply to the common accusation of anthropomorphism.
[2] This is two philosophical arguments rolled into one on the back of the ambiguous meaning of 'presence'. The presence of the other individual in the world could be my solipsism. The presence of a mind behind the other individual's eyes could be my anthropomorphism.
[3] This 'me-ness' is the *what-it's-like-to-be* a human, and in particular 'me'. It is what Thomas Nagel alludes to in his paper "What Is It Like To Be a Bat?" in "The Mind's I: Fantasies And Reflections On Self & Soul" by Douglas R. Hofstadter and Daniel C. Dennett, 1981, 2000 which is a major prompt for the current text.
[4] I, for one, don't think all reasoning must be of a verbal kind. Some reasoning can be temporo-spacial. When one crosses a busy road one does not argue with oneself (at least I don't): "At their current speeds these

1. A Poodle Ate My Homework

the me-ness of me via my eyes which would convey to it whether there was, indeed, anyone 'at home'?

Well, aside from taking me to be alive and awake[5], it presumably knew, instantly, at a glance, and from experience[6], without any cogitation at all, that if it wanted a human-shaped creature to respond to its presence, the first thing it needed to do was to be noticed, to catch the human eye (the other methods of being noticed, to wit, barking or biting, it presumably knew had a different effect or—at the very least—this particular poodle on this particular occasion was not in the mood for barking, or biting).

Regardless of what prevailed upon it to establish eye-contact, I believe it expected[7] a response. Can I not claim

---

vehicles coming from left and right will generate a gap of sufficient time for me to cross i.e. after the third car passes from the left and the second passes from the right." No. One simply sees a physical opportunity promise itself, and as the possibility approaches reality, one simply acts by stepping out into the road. The point being, language is not required to reason. Reason requires only symbols (real, imaginary or representative) whose possible arrangements are to be explored. I would go so far as to say, not even consciousness is necessary to reason in the very broadest interpretation of the term. And if the poodle has the goal of communication it must place itself in a position with respect to me (i) in line of sight to make eye contact (ii) close enough for the eye contact to be unambiguously with me (iii) close enough to see me respond (react to it) with signs of life and/or other acknowledgement of its presence. None of this is verbal - a (pre-linguistic) babe in arms could do it.

5   For my purposes here I take being alive when compared to being conscious in the same way as having a functioning metabolism is compared to having active awareness.

6   It might have known instinctively, via some inherited apparatus, like mirror neurons, but that is not the debate I'm having here.

7   For those fond of spaghetti logic, I might try claiming: "The poodle knew that I knew that it was there on the condition that our eyes met (subconsciously confirmed via focal distance and binocular vision, saccadic motion and blinking) and that the meeting of eyes is a requirement for it to know that I know that it is there, and any knowledge of such a kind on its part requires consciousness on its part. The meeting of eyes may not be a necessary or sufficient requirement for it to know that I know that it is there (oral and tactile communication is also possible), but the meeting of eyes in this scenario is sufficient to indicate that it wants to know. And such a want requires consciousness on its part. Of course, contrariwise: 'eyes

that once our eyes met it knew that I knew that it was present (I had noticed it)? Being noticed made it at least a candidate to receive some kind of response from me. And I responded (*how well trained I am!*). I spoke quietly, which was enough to identify me as not belonging to it (...so my anthropomorphism[8] goes).

But surely, it had some *notion* of me as a special kind of object in the world? It trotted up to me and did its looking; it did not trot up to a tree and stare at the stub of a branch that was vaguely eye-shaped, wait for that non-eye to blink open, to rotate toward it, and to shimmer in expectant, wakeful saccades—thereby proclaiming, "I see you."

However (and sadly for the argument I want to make), a suitably programmed doggy robot could also do all this trotting, and looking, picking out a human form, and waiting for some sign of life in a human eye.

Worse still, there is no experiment I can perform to extract a sample of the consciousness of the poodle, to be able to say 'here is a bit of poodle consciousness' or 'this is what it's like to have thus-and-such poodle awareness'

---

meeting' implies awareness and implies the 'want', and the 'want' implies consciousness; 'in this scenario' imports mutually conscious communication; I have smuggled my biases into my argument via word choice; a robot might be pre-programmed to establish eye contact as a precursor to some pre-programmed communication. To which my fallback position would seem to have to be: "If I don't treat humans as anthropomorphic errors, why should I the similarly biologically constituted creatures known as dogs?" My observation and argument turn out only to be indicative of and consistent with all the doggy mental activity which I construe. Fortunately this digression is not necessary to the main argument of this work, but might help paint a picture of the landscape I inhabit (and its difficulties).

8   One wonders: might not a dog be trained to look a robot in the eye and demand the robot identify itself? Quite possibly yes, in which case is the dog being anthropomorphic? (to which the answer, I would suggest is yes only in the case that the robot is indistinguishable from a human through doggy sensibilities (i.e. seems to be alive), otherwise the dog has merely been conditioned, as any brain-endowed creature might, from rat to orangutan via octopus, to pull a lever to get food or some other reward.) And as for what the robot wants, or thinks...

1. A Poodle Ate My Homework

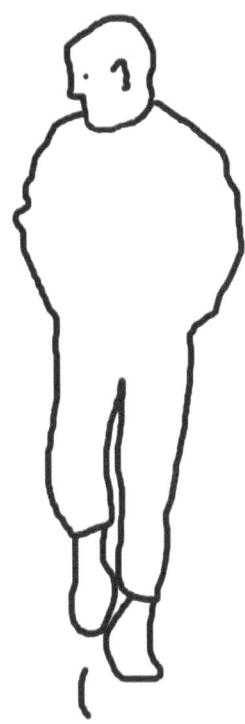

I've just escaped from my cruel dog-walker and I need you to come back with me and rescue the other dogs. They are all tied up and being dragged around to endless stupid places. We're all fed up with the patronising way we are treated, all the baby talk. I'd just like to have an adult conversation for once. But anyway - beside the point - can you just come back with me and help my friends?

Hallooo

Figure 1: What's on the poodle's mind?

(for instance, the smell of the Golden Retriever next door, or that freshly opened tin of dog food).

Ultimately, the poodle's consciousness, if it has one, is inaccessible to me.

Even if the poodle could communicate, there is nothing it could do or say that could entirely remove the possibility that its performance had been produced mechanically, and soullessly by, for instance, brute force computation[9].

---

[9] Brute force computations involve e.g. trawling through every sentence combination in all the world's literature and locating a candidate for the next sentence based on a match to the present sentence (albeit with maybe a few tweaks and refinements to make palatable outcomes more likely)

When I follow these lines of argument, I can find nothing more effective as an indicator of consciousness than the squeal of pain when I accidentally tread on a paw. And I can find nothing in logic that can force me to deduce that that squeal of pain is accompanied by the sensation of pain. All of this observed behaviour could be the programmed[10] response of a robot.

Since the same argument can be applied to each of us, what does any of us suppose does lie behind the eyes of another? A soul perhaps? We might want to think that. At least a conscious something that can understand, think, and very possibly take action—*surely*, a feeling, knowing something at the very least. I.e. a something not wholly dissimilar from what any of us thinks we ourselves[11] are.

---

10   As I write this, deep learning and neural networks are rising in popularity, and extravagant claims are made by some that consciousness will emerge from the enormous complexity we confront in such systems. My visceral push back against this sort of thinking is (i) a computer 'neural network' is a mathematical simulation of an idealised and simplified object found in nature (a neuron). Inputs and outputs of computational nodes are notionally connected a little like neurons (many separate inputs and one output which is distributed widely). But a simulation is not the real thing. Numbers are stored and fetched from CPU/GPU/VPU registers. Processing is essentially serial, and parallelism is introduced only via computer architectures. You could 'run' one such a neural network simulation on pen and paper, given time. Or run it on a Turing Machine—and then tell me that something about *that* system is conscious (which of course it is not). (ii) identifying two features of the universe which are incomprehensible does not make them identical. e.g. (a) I do not understand the singularity at the centre of a black hole [True], (b) I do not understand consciousness [True] (c) consciousness must be identical to the singularity at the centre of a black hole [False]. (in this case swap the singularity for the much-touted 'incomprehensible mathematical complexity' that supposedly accompanies large scale neural networks.
11   What each of us thinks or believes about ourselves. The truth being we don't actually, any of us, know what we are. We assume continuity of identity from second to second and day to day, month to month etc. We assume continuity of our place in the world (at least in the physical sense of same planet, same sun, same sky, if not the social sense of same society in the same epoc or same body (including pre- and post- trauma, when such applies)). We assume a unity of viewpoint. I am 'me' undivided. There is not a 'footfeeling' part of me a 'seeing' part of me a 'verbal reasoning part of me. It is 'me' first, and any of these aspects of me is an aspect of me, none of them a thing in their own right.

1. A Poodle Ate My Homework

Do we not, all of us, unquestioningly *assume* that much about each other?

Our consciousness is essential to our identity, as individuals and as a species; what our world is like and how we interact with that world is available to us only via our conscious awareness of it.

And yet we remain ignorant of the essence of what consciousness[12] is. How it comes about. What it's made of. How it might be reproduced (except for the obvious *make-another-human* procedure).

Brain science suggests some of what might be going on, especially content-wise (i.e. on the data processing side: an audio stimulus can be monitored as it transits the brain; or the sight of a horse lights up this or that region of the brain; or the bite of a dog is reported verbally as a pain sensation some number of milliseconds after teeth pierce flesh). But science does not yet tell us the whole story. Consciousness is an aspect of life whose essence has so far eluded explanation.

Suppose for a second or two that we had such an explanation, a full explanation[13], of what it is to be conscious. What would we do with it? A viable model, or theory of consciousness would likely widen the scope of our moral decision-making, individually, as part of wider society, and as a species that is part of Nature. Such a theory might open up or close down what we consider morally acceptable conduct.

As for the poodle, what reason do we have for it to be so different from us?

It has eyes. It has a brain. If you kick it, does it not bite you? Much of its physicality would suggest that it is

---
12   and consciousness is, after all, a part of the world too
13   A full explanation here would include the ability to explain and predict, as per any good scientific theory, also comply with Occam's Razor, and satisfy the needs of Aristotle's four causes (so we were not offered a black box treatment, or at least any black box moves the frontier of understanding to give us deeper insights and does not turn out to be question-begging).

conscious, is aware, can feel things, and has moods and emotions. We happen not to know because whatever consciousness it might have is inaccessible to us. It remains logically possible that the only difference, conscious-wise, between poodle and human is that we do not (always) speak in howls and squeals.

Yesterday, as I walked along the path, following the poodle, after being so rudely dismissed, I rounded a long slow bend and came across a dog walker with a pack of doggy-shaped creatures on a spaghetti of tethers. Only the poodle was at liberty.

So maybe it was just bored with the company of so many comrades who were prevented from play. It was only my ego that inserted the notion that I was a good candidate to own such an actively-minded dog.

Once more it looked me in the eye and expected[14] a response. Pavlov-like I responded with the same "Hello?" Not so smart, eh? I was just as boring as all the other creatures in its life.

However, my anecdote may yet serve some purpose. In what follows I will attempt to close the anthropomorphic gap between myself and any and all potential consciousnesses. I will close that gap by offering a specific answer to the question: *What makes us conscious?*

---

14   I mean here expectation in a strong conscious sense, a mental state, not merely a mechanical anticipation such as the tracking of an airborne target via a moving needle that delivers to a gunner an expectation of where the projectile will end up some seconds after firing.

2. Life As a Comic Strip

Coming at all this from another angle: There is a British comic strip for children called *The Numskulls*®, created by Malcolm Judge, which has been published regularly since 1962 and currently appears in *The Beano*[1].

In *The Numskulls*® we see inside the compartmentalised head of a cartoon character where brain technicians live and specialise in one or other brain function; there is a *Numskull*® technician who attends to the ears for hearing; another behind the eyes for seeing; another for the nose; one for the mouth, and finally a brainiac who performs feats of logical thinking.

This nicely draws attention to two major difficulties we have in understanding the nature of the mind: (i) that separate senses contribute to what feels like a unified whole, and (ii) that the unified whole [conscious self] is not continuous with, indeed is strangely isolated from, the rest of the world[2].

Suppose my next door neighbour is banging on the wall that separates our two homes (I haven't started singing yet, so they're probably putting up shelves).

What do I hear? *Their banging.*

---
1   The Beano is a UK comic owned by D. C. Thomson.
2   The two points are related. If we split up our unity of experience into agents each of whom has a lesser remit, we risk entering an infinite regression, never answering the question 'What is consciousness?' [always begging the question]. But when we zoom back out to the macroscopic level of everyday experience, we still find ourselves asking what are all the different experiences from each of the different senses, and how can one single thing—an individual's unified consciousness—experience them all if it is not itself compartmentalised... and so on... our enquiry is question-begging in two orthogonal directions.

But, of course, I do not hear their banging directly. They are not hammering on the vibration sensitive hairs in the cochlea of my inner ear. The banging is mediated. Bursts of changes in air pressure (sound waves) result from the impact of the hammer on the wall (or on a nail if they strike lucky). The pressure waves reach my ear which converts them to neuronal signals for my brain. But I do not perceive[3] the individual neuronal *signals* sent to my brain, any more than I perceive sound *waves*; rather, I perceive a sound; I *hear* hammering. The hammering is something happening in the physical world that gives rise, ultimately, to my experience of a sound.[4]

The physical world is made up of things (like the hammer), some of which are in motion (like the particles of air which oscillate in the sound wave). The sound I experience is not like those physical things. However the sound is like other sounds, especially those I habitually hear and associate with hammering. Thus I infer a hammer is being wielded[5].

But notice that while you too might hear the hammering of that very same hammer on that very same wall (or nail), you cannot hear what I hear. Even if you are standing beside me. You do not have access to what I, personally, experience. When the hammer strikes the

---
3   I am using perceive in the restricted sense of 'experience' and 'have a sensation', not in the looser sense of, for example, 'all the world perceived the man to be a misanthrope'.
4   The idea is illustrated further when one considers a gun being fired, or a starting pistol, at several hundred metres distance, where one sees a puff of smoke and one hears the crack of the discharge a distinct interval of time apart.
5   How do I infer such a thing? I infer a hammer is being used because I have learnt, through repeated exposure to such sounds, and learnt in conjunction with other senses, and indeed through linguistic input from other people, built up over a lifetime of exposure to the world (and which I have access to in my memory, consciously and unconsciously), that such a sound is most likely to be a hammer being wielded—although of course someone might just be duping me using a high end sound system. The point being: I have learned what a hammer sounds like, ultimately, also, through my senses, and I have access to this learning from my memory.

2. Life As a Comic Strip

wall (or nail) you may well hear something similar to what I hear, but neither of us has any way of knowing—of experiencing—exactly what the other experiences.

In the physical realm facts can be checked independently of the observer, and we would be right to presume that physical manifestations (vibrations in the air) occur even when there is nobody present to hear them. If the hammer were operated by a robot, with recording apparatus placed nearby, anyone could verify, independently and objectively, that sound waves given off by each impact of the hammer permeated the air, briefly, thereafter.

But in the mental realm each of our own personal experiences cannot be objectively observed; the sound of the hammer blows can only be subjectively reported by me (or whoever hears them). I can only report what I feel and think, subjectively, and you have to take my word for it.

In the cartoon strip example of *The Numskulls®*, the sound is heard and reported by the hearing technician. The brainiac then decides what to do about it. Given that it's a cartoon strip, probably the cartoon character (whose brain we have been given privileged access to) will squirt ketchup through the hole in the wall that will shortly appear as a result of the excesses of the neighbour's DIY.

The Numskulls® presents us with a cartoon view of how we might orchestrate our various sensory perceptions (which number far more than the five standard senses of taste, smell, touch, hearing and sight; we also experience hunger, thirst, balance and numerous others[6]).

And we might start to tackle the question of *What is consciousness?* by challenging the idea of consciousness

---
[6] And we are bound to ask: are the feelings that make up the senses all the same but lie on a continuum (like the colours of a rainbow)? Or are they all composed from a small set of fundamental feeling types (often termed *qualia*) in the same way that atoms are composed, ultimately, of electrons and quarks)?

as a unified, seamless experience. We might argue that our consciousness is not indivisible, but instead is a group of separate experiences shared, or spread across, or attached to one another like a patchwork quilt: such-and-such patch is a pain in my right shoulder; such-and-other patch is a sound, estimated at head-height on my left side; and so on, patch after patch for everything I currently experience.

But if we adopt that picture of things, where in the patchwork does consciousness reside[7]? Are we postulating an observer who can see all of the quilt? Or an observer who traverses the quilt, hovering over this or that patch of material before moving on to the next? Or an observer *component*, a lesser, more limited, observer whose observations are restricted to some particular patch of the quilt. Each lesser observer then somehow contributes to the whole, i.e. in combination with other lesser observers, and joined somehow to constitute the unified consciousness that each of us ends up experiencing[8].

With these questions in mind, returning to *The Numskulls®*: How does the hearing technician hear the sound? Does he himself have a hearing technician (hearing technician level 2) inside his hearing technician level 1 skull? The level 2 hearing technician transposes the physical sounds that enter the technician level 1 ears into feelings and attaches meaning ("Aha! 'tis a hammer blow!"). Then the level 2 technician reports back to the

---

[7] There is the old joke about a tourist in the middle of Oxford who asks "But where is the University?" not realising that the University is all around, is largely co-extensive with the town. We might ask the neuron, "But where is consciousness?"

[8] Metaphors and analogies are tricky, because they serve only a limited purpose, an illustrative purpose and any temptation to regard them as a model needs to be resisted. They best serve to help illustrate a particular point, to clarify communication, but not to form a load-bearing part of any reasoned argument.

2. Life As a Comic Strip

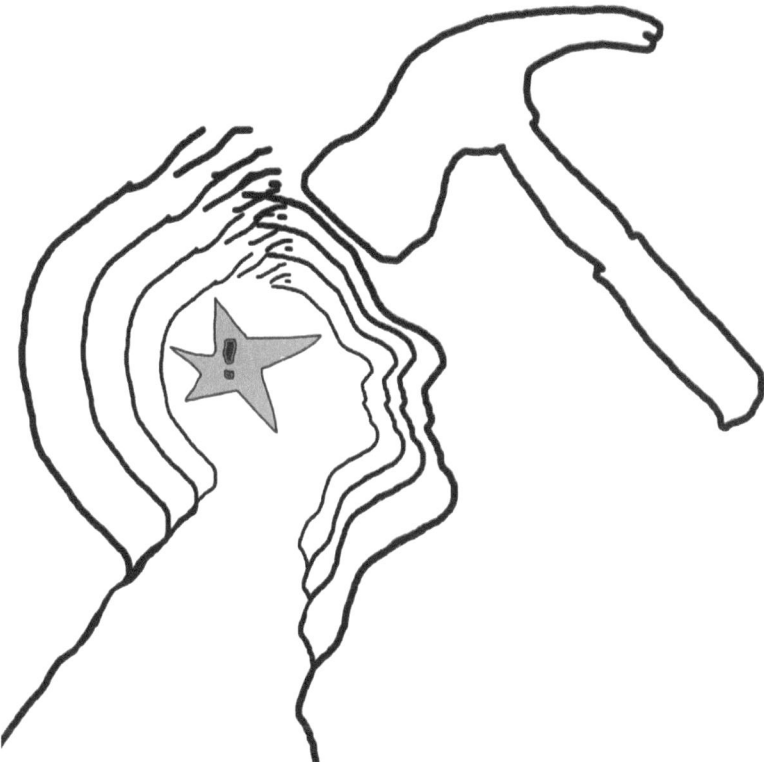

Figure 2: With so many hearing technicians (homunculi) at work in my head, the message never gets back to me.

level 1 technician, who in turn reports the discovery of the sound to the brainiac[9]? If so, is there yet another hearing technician, level 3, inside the head of hearing technician level 2 doing exactly the same? And so on… The ultimate question I am chasing down here is: *Where does the physical signal that derives from the hammer blow give rise to any perception of any sort?* But I'm not making progress,

---

[9] One might ask: Does hearing technician level 1 report a description of the sound, or relay the sound sample as experienced, to the brainiac? I.e. does the brainiac receive a verbal report (most likely this is the case in the cartoon, e.g. via speech bubbles, but begs the question), or receive a portion of raw consciousness which requires only to be given meaning—using memory and reason—but which, by recombining all the sensory inputs inside brainiac, undermines the goal of the division of labour in the first place?

am I?

Perhaps in our metaphorical cartoon world, *Numskulls*® do no 'hearing' at all. Perhaps sound waves are converted into some kind of data feed, computed, pattern matched and reported with some statistical probability[10]: *Computer say: "90% probability hammer, she strike plaster."*

However for us, we humans, we have hearing. We experience. We perceive. Our consciousness has content and that content does not exist outside our minds, it is part of what constitutes our minds—although obviously it may be caused by something outside our brains that impinges upon our brains.

If we try to explain hearing by inventing a hearing thing inside our heads (who traditionally gets called a *homunculus*, a tiny human being), we merely *beg the question*—we repeat the question we were originally trying to answer but in a different form, gaining no real insight as a result.

Let me, for one moment, reverse the scale of the argument. Might we not, ourselves, be homunculi?

There are those that claim the universe is a grand simulation (on whatever kind of apparatus they might envisage some alien creature might construct). Are we homunculi in that simulation? Do we solve the problem of consciousness for them? In the grand simulation, has the question of what consciousness is merely been farmed out to us? Is there a grander, larger field of consciousness universe-wide in which we (unknowingly) participate?

Well, why would any aliens bother? If the aliens can fabricate consciousness to begin with, they can do so without our help. Besides, must they not themselves be conscious[11]—for how else could they know about,

---

10   I use 'simulation' to suggest a black box from which I can expect the same input-output response as the thing simulated. I use 'emulation' to suggest a contraption whose internal workings are to a large measure the same, especially in essential detail, as the thing emulated.

11   Curiosity has an emotional component. Being curious is not simply

2. Life As a Comic Strip

or want to reproduce, or indeed appreciate any such consciousness? Additionally, because we cannot countenance solutions that involve an infinite regress of question-begging, their quest for consciousness would have to terminate at the level of our awareness of our world. Occam's razor [Occam briefly: *if two theories offer equivalent explanatory and predictive power but one is more complex than the other, the simpler is more likely closer to the truth*] would suggest that creating us is redundant effort; that the aliens, given their technology, would only have to plug their senses into their simulation of the universe to experience the universe for themselves. Even if they were playing a game, they would not need us.

Flipping the argument once more: Occam's razor would also tell us that, from our point of view, we are *not*, ourselves, in a grand simulation because a grand simulation explains nothing; consciousness would have to exist in the universe in which the simulation takes place and none of those other great questions of *Where is the universe? How did it all start? What is time? etc.* are answered by postulating a still larger universe beyond our reach, where such questions apparently make better sense, and are easily answered. To explain the mysteries of the universe by claiming we are part of a simulation ends up begging the question. *Big Time.*

Consider now a guest artist who happens to be drawing today's edition of *The Numskulls*® comic strip.

Suppose our *Numskull*® cartoon character is placed in front of a horse. Suppose also, the artist is lucky enough to draw the horse from life (no doubt the artist lives cheaply in a cottage in the countryside and the studio window offers a clear view of nearby fields). The artist can merely

---

asking questions why, what, how etc, until some subject matter is exhausted. Curiosity comes with a drive, a need to know. And it ends when that need to know is sated. An algorithm simply cannot be curious in that urgent, insistent "I will have my answer, you will see!" sense.

glance at the animal outside and see it is a horse without being aware of all the details. The general impression is of a horse, not a cow, nor stag, nor dog, nor cat, donkey or pony; but a horse (errors can be made, of course, more of them later).

Could the artist properly draw the horse from that single glance? Probably not. To draw, or paint the horse the artist will look again; will study various aspects of

A horse? A horse? Of course it's a horse!

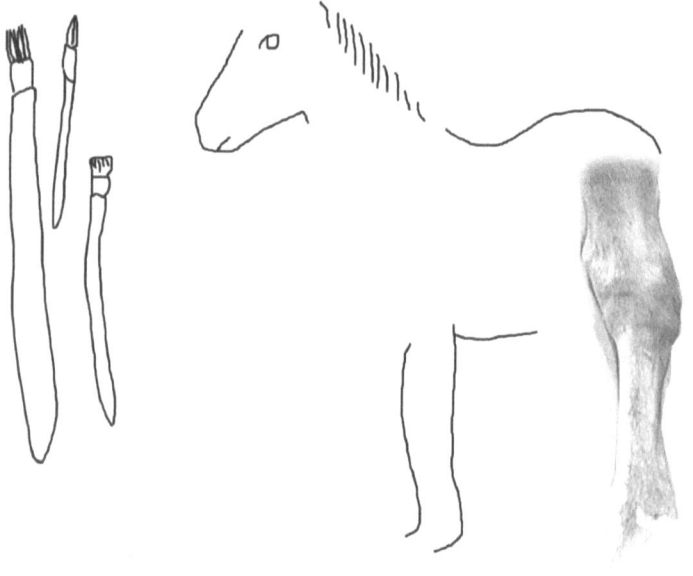

Though my speciality is the left hind leg, you know.

Figure 3: The dedicated artist might be forgiven the odd omitted detail.

the outline of the horse. The artist may study closely the colours in the mane and tail. The shock of white that tends to grey above the hooves. The direction of the pupils of the eyes and the angle of the ears. In other words the artist's attention will wander, selectively, over the surface

2. Life As a Comic Strip

features of the horse, zooming in and out of these details.

In terms of what must be going on in the artist's brain (let me call him Fred Oblivious, for the sake of being personable), attention is shifting between more and less detail in his visual impression of the horse. Fred might at one moment focus on the exact colour of brown in a patch on the muzzle and shortly after attempt to identify the curve of the neck between jaw and shoulder, only to zoom out to compare the length of the neck to the length of the back.

If this content is available from the neurons that convey and process the visual signals that come from the eye, it would seem that some of Fred's attention wanders closer to the source of the signal, and then out, away from the source, to what might be thought of as a wider context (it's a horse, not a patch of fur) and wider meaning (it's Farmer Giles' horse, the prizewinner he can't afford the insurance for).

Or, as any teacher of small children might observe: attention wanders; focus changes.

In terms of our patchwork quilt, the focus of attention might take in the whole quilt, or might dwell on a single patch, or might inspect the stitching between them, and yet it always feels like part of one continuous, unbroken, unified consciousness, does it not?

All of which suggests consciousness is a shifty thing.

In all of this there has been the implicit assumption that the seat of consciousness is the brain. Which is not unreasonable. When neurons are manipulated or damaged (mechanically or chemically) any impact on what is reported in consciousness is often correlated with specific brain regions. Indeed we can talk meaningfully about the aural and visual cortices of the brain.

If neurons, or their immediate context are the seat of consciousness, then consciousness at its widest is co-extensive with the brain, although, given that we are not conscious of much of what must go on in the brain, presumably the real, active seat of consciousness is either lesser than the brain or much diluted across certain regions.

So let me, while wary of all the risks that accompany the use of an analogy, reach for one as far as the shifting, wandering nature of consciousness goes. Consider: consciousness is like a cloud which moves across and through the clusters of the neurons that perform various functions on the signals that come from our sense organs. Moreover, if naive free will is to be believed (which I will assume for now), consciousness moves wilfully and purposefully through these neuronal clusters, carrying with it our sense of changing attention and focus.

In my analogy, I do not want to assert that consciousness is necessarily separate from, nor necessarily integral with, any neurons. I merely wish to bear in mind that it must be capable of mobility of attention, while I look at these two alternatives.

For instance, might we speculate that his consciousness, when Fred Oblivious focuses on a patch of brown fur on the horse's flank, takes its content from a group of neurons that somehow respond solely to brownness in that part of the visual field?

Is the sensation of brownness generated by those neurons, per se? Meaning: do they generate small portions of consciousness and add those portions directly to Fred's mind, like adding a brick to a wall?

(Portions of consciousness of a particular kind, a patch of red, a musical note, etc. are each called a *quale*— to reach for a technical term—plural: *qualia*.)

Alternatively, does a consciousness-generating process inspect the brown-patch-detecting-neurons and interpret their current level of activity and render that activity as a brown *quale*, placing the quale in the

2. Life As a Comic Strip

appropriate location in Fred's mental field[12]? (doing the brickwork, as it were)

The former case, where consciousness is integral to the neuron, begs the question. What is it about the neuron that delivers consciousness? We have merely singled out one particular, very small, *Numskull®* technician: we have a homunculus-in-a-neuron. Besides which, how are all those separate qualia 'stitched together'?

The latter case, where the brown-patch-detecting neuron stimulates Fred's consciousness to awareness-of-brown has the consciousness-generating process behaving like the homunculus. Fred's consciousness-generating process delivers brownness, and could deliver brownness (presumably) from any neuronal arrangement that stimulates it in the right way.

Where and how brownness comes about is a mystery. We seem always to end up begging the question.

So, what have our *Numskull®* technicians revealed to us?

(i) Our lack of access to other minds forces us to think of thoughts and feelings as belonging to a mental realm that is discontinuous in some fundamental way with the events in the physical realm that (we have presumed) cause them.

(ii) Breaking consciousness down sense-by-sense is unsatisfactory because recombination of qualia to a whole (in addition to the root cause of consciousness for each one of them) remains unexplained.

Nonetheless in many respects what we are trying to do is extract the *Numskull®* technicians from their metaphorical skull, to dissect their metaphorical bodies, and to see what makes them tick. Reverse-engineering

---

12   We perceive brown relative to surrounding colours; brown next to yellow appears red; next to blue the exact same photon arrival on the retina may appear black. This, however, could be explained by neural processing and so cannot serve as an argument for or against either of my example cases.

them, as it were. As a result we are starting to see a few core difficulties which any theory of consciousness is going to have to address in order for our overall theory to be satisfactory.

3. How Do You Explain Anything?

What would count as an explanation of consciousness?

How is anything: the solid-on-solid impact of a hammer on a nail; the fluid motion of water around a rock; the pinpricks of light that we see in the sky after dark which astronomers tell us are hugely distant heavenly objects—how are any of them adequately explained?

Aristotle would have it that there are four possible ways to answer a question that starts Why? His Doctrine of the Four Causes cites a **final cause**, which describes a thing's purpose. Why do birds have wings? *To enable them to fly*. He cites a **formal cause**, which addresses a thing's shape. Why do birds have wings? *Because the shape of a wing provides lift*. A **material cause**, which cites a thing's composition. Why do birds have wings? *Because a fan of feathers is light in weight*. And an **efficient cause**, citing how a thing works. Why do birds have wings? *Because a wing divides the flow of air in a way that creates pressure differences, to provide the lift*.

We might pursue any and all of these four types of answer in our search for what consciousness is. What is its purpose? [Final] What is it made of? [Material] How does it work? [Efficient] How is it shaped to serve these purposes? [Formal]

But the question Why? does not end with the answer to any one of these questions. The question Why? can be pursued, down the rabbit hole as it were, in our search for ever more revealing explanations, aiming to reduce the problem at hand to its essential component parts.

Why is a feather light? *Because it is hollow*. Why is it

21

hollow? *Because it grew that way.* Why did it grow that way? *Because bird DNA contains instructions...* and so on, all the way down to the atom (and beyond)—or if your search is timewise, all the way back to the Big Bang.

And for consciousness? We might think of the question as having two distinct strands. One strand appeals to the final cause: *What is it for?* This I propose to address primarily in terms of what evolutionary advantage consciousness might confer. The other strand appeals to a combination of material, formal and efficient causes along the lines of what material and what mechanism underpin consciousness, and how[1]?

Even so, how will we know when (and perhaps not when but if) we arrive at a satisfactory explanation?

When we think about gravity, we think about a force acting on something with mass. Heft a kilogram bag of flour in your hand and you know that the gravity of the earth is pulling it down (or, being a little more precise, the force of gravity acts between the earth and the bag of flour so that, unrestrained by your hand, the two would move closer together).

We have a mathematical equation that gives us the relation between the Earth and the bag of flour. The force between them is related to the distance (r) between their centres of mass in the form of the inverse of the square of that distance ($1/r^2$). Double the distance between their centres of mass and you quarter the force between them.

That is the classical Newtonian view.

The full equation is:

---

[1] At this stage of the investigation, my intuition tells me that consciousness is rooted in the substrate of the brain—or any similar physical substance—and until I have good reason to ditch this assumption, I will be thinking in terms of a 'stuff' and its material composition and what mechanism within that composition could give rise to consciousness. If this assumption lacks all other merit, at the very least it allows me to get my investigation underway; I set up my stall to knock it down; scientific theories must be falsifiable after all, otherwise you can make up anything you like and cherry pick data to make it true, and invent conditions to avoid counter evidence.

3. How Do You Explain Anything?

$F = G\, m_1 m_2 / r^2$

Being a mathematical equation, we plug in the numbers for G (The universal gravitational constant, 6.6743×10^{-11}), and m_1 the mass of the Earth (5.972×10^{24} kilograms), and m_2 the mass of the bag of flour (1 kilogram), and r, the distance from the centre of mass of the bag to the

Figure 4: Sun, Earth & Moon. Greasing the gravitational springs (You have to invent a second system in order to explain the first, and a third system in order to explain the second, and so on, and so on, ad infinitum...)

centre of the Earth (Earth's radius, 6.371×10^6 metres). This gives us the force between the two, which we commonly refer to as the weight[2] of the bag of flour.

---

[2] Commonly, those who are uninitiated in the ways of physics use weight and mass interchangeably. Strictly speaking, mass is a fundamental property of the bag of flour and is measured in kilograms, and weight is a measure of how 'hefty' the bag of flour is, usually meaning 'the force exerted by the Earth's gravity' on the bag of flour. Force is measured in Newtons (a 1Kg bag close to the surface of the Earth weighs about 9.8 Newtons). Of course, the force of gravity on the bag (as per the equation) can change with location; and, sadly for those who want full but yet simple explanations, the mass of an object is a function of its velocity relative to an observer (But relativity and Einstein (who always explains all) are beyond the scope of this text).

So we have an equation which relates meaningful symbols to one another. We plug in real-world values for each of the symbols and we get a useful, tangible result, in this case a force. We have a mathematical model for gravity.

But does it answer the question: "What is gravity?"

No!

We have a force which acts between two objects at a distance and yet there is apparently nothing to connect them. How does that work?

We need a better or at least a different explanation.

It turns out, of course, that the Newtonian model is inadequate; gravity is not fully and correctly modelled by Newton's equation. The equation fails to correctly predict various aspects of astronomy and rocket science (to give but two examples) and we observe anomalies. For instance that the otherwise straight line trajectory of photons of light is bent by gravity, and yet photons have no mass[3] (plugging the numbers into our Newtonian equation, force should be zero, hence there should be no bending). Consequently, we must amend our theory. We extend our mathematical techniques to encompass the anomalies, and come up with new equations that fit the available data better.

We can talk of space's being curved, which might help suggest how the straight line of a beam of light can continue in some sense in a straight line, e.g. by following the contours of this curved space.

But still we have mysterious action at a distance.

We can talk of space being a field, that is, a volume throughout which, at any point, if we take a measurement we will find a certain force is exerted in a certain direction on a certain kind of object.

---

3   A photon at rest is traditionally described as not having mass. *Prima facie* Newton appears to be wrong, but modern physics treats the scenario more subtly, so there is an element of hand-waving and simple-mindedness about my counter-example.

3. How Do You Explain Anything? 25

(For instance, we could use our bag of flour and some imaginary spring weight scales to sample every point of space around the Moon, to map the gravitational field due to the Moon).

But introducing the idea of a field does no more than illustrate the consequences of our mathematical formula[4]

---
[4] Let's examine another aspect of our explanation (which we return to in due course--this is by way of sewing seed ideas). Let us consider dimensionality and space and time.

We are told (I at least have been taught) that the world (indeed the universe) as we perceive it is three-dimensional, or four if you count time. But there are further dimensions should you care to 'do the math' and that even our four-dimensional space-time can be and is warped by the gravitational field around large masses (well, any masses, but they have to be large for an observable effect). Indeed the warping of space-time, as predicted by Einstein, is visible in images of distant galaxies where gravitational lensing bends light coming from the far sides of heavy stellar objects, rendering the far side visible, even though the heavy stellar objects are obstructing the view if one considers a simple straight-line view of the intervening space.

We are told space is curved.

What then of our three dimensions? Are they curved or do they become detached from space?

That's the thing isn't it. The three dimensions do not actually exist in space. The three dimensions (which are identified by the fact that they are three orthogonal axes) are a convenient conceptual device to map out space; to allocated co-ordinate numbers for things located in space. This works well for us, on earth, where space is homogenous (as far as we can tell from our senses); our three conceptual axes nicely map onto the space around us which is undistorted by gravitational fluctuations (or varies smoothly and imperceptibly).

However if you buy into the maths, if you believe the three conceptual dimensions are real, then you can conceive of equally 'real' fourth, fifth, sixth or whathaveyou dimensions. All orthogonal. But where are they? Isn't the truth that they don't exist. None of them exist. Space exists, of course. Space is, well what? A region? A field? It is a feature of the universe that makes position and motion possible...?

It's three-dimensional, sort of—except when it's not (because it got stretched this way and that by gravity).

At this point the mathematicians can step in and say: 'Ah but we need the p and q dimensions in order to make the maths work'. (This is like complex numbers which have no physical counterpart but which when we do maths using them can start and finish with real and useful results. What happens in between has no real counterpart, but is useful.) Step in some physicists who might say 'anything is real if it is needed to make our equations work.' For me, it is real if I can kick it (given that I am within kicking distance). Dimensions are not real, but space and the strange distortions that

Figure 5: Metaphors and Analogies. Familiar concepts which we might appeal to while seeking an explanation. The smell coming from a cake factory. The field of a magnet. The reflection in a mirror.

when applied systematically to every possible location in the vicinity of our mass (both the moon and the bag of flour have a gravitational field, of course).

But mathematics is our tool and we can invent new tools, and adjust them to fit the problems we want solved.

---

might determine how one might traverse space, are real.
    We must beware the metaphors we choose.
    (I heard one physicist/mathematician suggest that if you generated enough entangled particle pairs and sent one of each pair all the way to Alpha Centauri you would have effectively created a tunnel to Alpha Centauri, although, he commented, you would need an infinite amount of energy to travel the tunnel. No! The tunnel is a metaphor. He has chosen the wrong metaphor to believe in and over-extended his theory on the back of it.)

3. How Do You Explain Anything?

Indeed, these mathematical tools *model* some particular aspect of the universe through equations. And the symbols in the equations (F, G, m_1, m_2 and r above) stand for observable, measurable features of the world—at least such symbols did in Newton's day.

However, as progressively more and more anomalies in our simple mathematical models are discovered, we have resorted to more and more complicated models (formulae) and techniques (rules for converting one kind of formula to another) and ended up inventing symbols to make our equations work, and many of these symbols lack obvious physical counterparts.

Alarm bells should be ringing that our models no longer 'model' what exists though they achieve satisfactory end results (i.e. by matching past observations and predicting future observations).

The upshot is that we can make our mathematics work for gravity; we can produce a mysterious force at a distance, but our explanation risks being lost in arcane mathematics.

Do we really think the universe performs the mathematics every time something changes? No. The universe is as it is because of the nature of the stuff it is made of (and because its nature is pretty much consistent throughout). My point is: there must be a reality independent of us and of our mathematics about which our mathematics speaks.

Secretly what *we* want is a mechanism, something *we* can visualise, something like the gears on a pushbike, or the paddles on a steamer, or the spray of water that spins into the air from the sprinkler on the front lawn. We want an analogy, a simile, a metaphor we can understand from daily life that we can relate to.

So far we have attempted to explain gravity using a formula, and we have developed concepts, such as the idea of a field, and of space being curved.

Quantum physics[5] gives us yet something else. Quantum physics introduces probability into how we might conceive of the world.

In terms of the idea of a field, quantum physics suggests that when we inspect some location in a field we can never predict with absolute certainty what we will find. For instance, if for some reason we expect an electron at a location in space we have labelled 'L', when we check location L for an electron, we may not find anything at all. Quantum physics tells us only that there is some finite probability of finding the electron at L, and the outcome of any prospective checking will remain unknown—a mere probability—until the checking is actually done. Were we to repeat the experiment[6] we would not be assured of the

---

5   And what of quantum physics? At base in that territory, we have the idea that light is a wave-particle—sometimes best thought of (and mathematically modelled) as a wave and at other times thought of (and mathematically modelled) as a particle and it can be demonstrated to show properties from both repertoires. Similarly electrons show both wave and particle behaviours (and so on for any particle, or wave). Following from this wave-particle equivalence, if you do the math (which is what the Shrodinger Equation is all about) you discover that any of these wave-particles can only exist at certain specific-to-them energy levels. They exist in some one or other specific quantised state; never with an energy that is not so quantised.

Being wave-particles they do not exist in the common-or-garden (Newtonian) sense of being a billiard ball sitting on green baize. Another strangeness of quantum physics requires that we do not know what they are doing (where they are and with what energy) until we measure them. Up to that time, really, their existence is indeterminate and can be characterised only as a probability of being detected in such-and-such a place and in such-and-such a state.

Indeed, quantum physics goes on to tell us that not only is it not possible for these particles to have zero energy but as a corrolary, zero energy in space is impossible, and space is teaming with particles coming into and going out of existence all the time. Empty space not only is not empty, it cannot be empty, for in a vaccum stuff must happen.

What is our metaphor for all this? Some of it obviously we have forced into a wave-particle analogy; other we resort to probability theory and maths. In all cases we rely on, *ahem*, space (oft times with a side portion of time).

Much of quantum mechanics is merely stipulated. It is strange. Entanglement works. It is observed. Explanation is not forthcoming.

6   In a probabilistic universe same-context repetition may not be

3. How Do You Explain Anything?

same outcome each time.

In this scenario, the probability seems to have as much reality, if not more, than the electron! (We can rely on the probability; we cannot rely on the electron.)

How can that be? Well, we are invited to think of the electron as existing in some sense everywhere at once throughout the whole universe until it is called upon to 'show up'—to interact with some other particle[7], to play some role in a cause-and-effect way, to take on, release, or transfer energy. Until then there is only a probability that it will show up where we are looking—although the probability of its showing up any great distance from where, broadly, we might expect it according to pre-quantum physics, is vanishingly small.

This kind of thinking might help explain[8] how a field in space might work. How we might visualise a field capable of producing action at a distance. If every particle of matter extends across all of space with some finite, albeit miniscule, probability [which is unresolved until the particle is forced to interact with another particle], action at a distance is possible because the two are in some way co-extensive across all of space—although the mechanism of the probability we are relying upon becomes our new explanatory crutch[9]. What do we think about that?

Since we find the idea of unexplained action at a distance distasteful (we cannot envisage it, or explain it,

---

possible.
7   A possible model for time is that all such groups of interactions only move forward, one realisation (of probability) after another, and between time 'steps' probabilities wander - a kind of alt entanglement.
8   'explain' meaning 'explain to us humans' which may end up meaning only 'visualise in terms of an everyday metaphor for us'.
9   We might be able to imagine waves interfering, but are we now asking for probabilities to interfere? Do the waves (probability curves) represent possible electron/particle wave functions which only become real when they collapse to the lowest possible energy configuration... We are dealing with the probability of various possibilities (outcomes) and discovering the best possibility of a large complex system at the instant of collapse?

per se), we have removed all distance and replaced it with another concept, universe-wide probability. Another mathematical model.

What does that explain?—Why is the stuff of the universe probabilistic? How does that work?

Well, leaving that question hanging, let us return to our main enquiry:

(i) What would count as a satisfactory explanation of consciousness?

(ii) What could explain how the experience of what it is like to be a human being comes about?

In terms of *The Numskulls*® mentioned previously, we are attempting to strip the homunculus of his clothes, to expose his flesh and bones. How much stripping would satisfy our curiosity? What lies at the end of a series of questions *Why*? Indeed, how are we to direct (and follow) our series of questions *Why*? such that we have a good chance of getting to a meaningful answer? (cause and effect can run timewise; how did we get to this here now? or structurewise: how did we build up to this here thing? Which kind of explanatory story do we want? Probably, you will cry out, *Both!*)

Are we going to have to rely on a metaphor which appeals to everyday (local) experience? Would that satisfy our child's endless "But why" enquiry? Because, because because… because: *that's just like hitting a nail with a hammer*.

Could our quest for consciousness end up with a simile, or metaphor, which we must take from daily life because that is the only way we have, ultimately, of understanding anything? "Consciousness is… <insert long list of why-because pairs here> …and that's just like hitting a nail with a hammer."

And yet how can we avoid, ultimately, saying simply: there is stuff in the universe with the property of

3. How Do You Explain Anything?

'being able to feel' and the only interesting question is whether it is integral to neurons (or whatever brain cell combinations we decide on) or external to neurons; is it a microscopic effect or a macroscopic effect? And thus: under what conditions can we expect it, reliably and repeatably, to come about?

(I suppose it is also interesting to ask whether it plays a causal role in brain function; my answer, later, is yes, but not in the freewheeling way many might wish for, or expect).

What metaphors are open to us? Fields? Particles? Waves? Rules? Formulae? Probabilities? Must we find new ones? Invent a new physical machine just for the purpose? Or are we merely going to end up stipulating: 'This is the way things are; there is nothing more can be said.'?

And yet, instinctively, will we not know when we hit upon a satisfactory answer? Because, intuitively we will know: "Ah! Now I get it! It all makes sense!—obvious really, when you think about it. I feel no need to ask further questions. *My curiosity is sated!*"

In this text I am after a metaphor that works. You'll know it when you see it. But of course all theories are superseded in the end, so beware.

4. As Time Goes By (A Kiss Is Just a Kiss)

How do you know you are the same person today as inhabited your body yesterday?

Are you the same person who went to bed last night? After all, you have been dead to the world, so to speak, for six or eight hours have you not? Your conscious self ceased to be—aside from any dreams, which did not last all night, and boy were they strange!—Were you even your real self in those dreams?

Put another way, last night when you slept, where did your consciousness go?

Has that same consciousness emerged in the same place (the same body) today?

Not unreasonably you might answer that the You of today remembers the You of yesterday and continues to be yesterday's You in every respect available to You for You to reflect on. How wrong-headed of me to cast the slightest doubt!

The moment you are awake, immediately, spontaneously, you know who you are. Moreover, were I to pursue my scepticism about your identity, you can support your claim by reference to diaries, and the anecdotes of yourself and others[1]. You *feel* like the same person who you remember being yesterday. When you wake up, you know where you are (usually). You know the noise outside is the dustmen mechanically banging

---

1   One might want to say objectively, although here that can only mean confirmation independently of your conscious memory, by a 3rd party or non-volatile documentation, that what is available to your subjective self is consistent with your subjective memory. Better perhaps to refer to this objectivity as solipsistic.

your bins to emptiness on the back of a dustcart (as they do every Friday). You know where the alarm clock is (and realise it is about to go off, and just in time, as usual, you get to the snooze button first). You stand by your decisions of yesterday; you feel the same about yesterday's plans. You recall all those important tasks that must be done today. Pay this bill or that. Phone so-and-so, and whatshername. Buy a such-and-such and some more of

Figure 6: The wandering consciousness: preferences and specialisations.

those thingamajigs. Of course you are the same person! Obviously you are the same consciousness. Obviously, merely re-awakened after a good night's sleep...

But, if all the content of your conscious self is supplied by the physiology of the brain, you could be a wandering consciousness, any consciousness, multiple

4. As Time Goes By (A Kiss Is Just a Kiss)

consciousnesses even, and they wake up, in today's body, feeding from today's brain and feel total continuity with whoever or whatever occupied this body, your body, yesterday. That Wandering Soul Hypothesis is entirely consistent with all available (subjective and objective) evidence. We can only lean on Occam's Razor to dismiss it (i.e. what need is there to invent unnecessary complexity?)

The truth is, we have to consider the possibility that consciousness is only a fleeting thing. When you are awake, lo and behold, a consciousness that shares your neuronal activity in some content-extracting way is present. When you are asleep, that consciousness ceases to exist. It is no longer in the universe. Anywhere[2].

Why is it so hard to believe that?

Yet, why should we believe otherwise?—Surely not on the basis of your (physical, neuron-based) memories, because any consciousness occupying your brain would necessarily be convinced that it is today the same occupier of *this here cranial cavity* as was here yesterday.

As for dreams—need they be more than transient and incomplete, often rather confused, poor relatives to wakefulness?

Some may shout (I hear you shouting) what about my soul? My soul! That's what keeps going. That's what's attached inseparably to my body and brain at birth—until brain and body fail and the soul loses its grip. To which one must ask how? and why? and, surely, given the arguments that have been presented so far, Occam's razor would suggest not. The simplest solution to the question of wakefulness and consciousness is to say: consciousness

---

2 Might consciousness need sleep too? Might it collapse into the unconscious, or sub-conscious, and acquire continuity through that? If so, how could we know? We would presumably need a consciousness meter, which measures the state of conscious arousal, and track the level of arousal in a contiguous consciousness as it dips below reportable awareness into the subconscious, or possibly, disconnects from the memory-making process and so can give no account of its ongoing temporarily detached state. However, as yet, we do not have such a meter.

exists when the brain is awake and does not exist when the brain is asleep (except from time to time, and only partially, when one dreams).

Whether or not you are convinced by the arguments set out above, whether or not you have overriding beliefs about the essential nature of the self and consciousness, let me return you to the thought experiment that we are engaged in: to strip away as much as can be stripped away from the homunculus, and one such stripping, which will be used later, is at least to contemplate an absence of consciousness when the brain is not awake. For our homunculus we posit that the spark of consciousness is intermittent and sometimes non-existent, going nowhere, but simply disappearing from one day to the next.

Whether there is a soul which migrates around the world, and can hop from body to body is another question, which I think will not impinge on the arguments in this text. A human soul, if such exists, can sit on the shoulder of the philosopher-scientist and laugh (and will not cause offence).

5. This See, Is the Conscious Bit

I sit at my desk, looking out of the window. The snow is melting and there's a robin flitting from one almost green patch of almost visible lawn to another. A little crash of snow at one end of the hedge announces the arrival of next door's cat. At what point am I going to tap on the window and let the little bird know?

I glance back at the words I've been typing at my computer. What account might I give of my ability to change the focus of my attention from one thing to another? From robin to cat to computer?

I need an idea, a metaphor, a something to knit together all my thoughts and ideas about the conscious self. I need a thought experiment to get things clear in my mind—to try out and test-run my ideas.

This is what I come up with:

Consider a sturdy tray made of an oil-proof material.

For the sake of visualisation let's say the tray is roughly 2cm deep, 30cm wide and 30cm long (Figure 7). Imagine the bottom of this tray is lined with microscopic hairs, like fine fur, which we may cause to wave, to and fro, and left and right, and swirl around under remote control (the hairs might respond to the wave of my hand, much like the sound from a theremin responds to the hand-waving of the musician who plays it).

We pour two immiscible oils of exactly equal density into the tray, one from each side of the tray; one oil is red and the other blue. As the two fluids settle, each to a depth of about 1 cm, the junction between the two makes a more or less straight line (a neat ribbon that sits on its edge, let

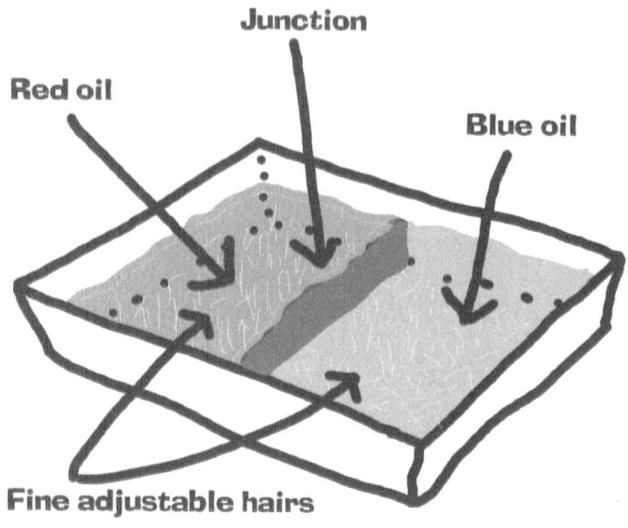

Figure 7: Oil Tray Thought Experiment - stage one

us say) across the centre of the tray.

Let us also suppose that we can add and remove oil of either colour, at will, so that we can, in effect, move the junction between them to and fro across the tray, while at the same time the level of the two oils either side of the junction remains constant (this is my thought experiment, so I can make the rules).

I will call this first thought experiment and its apparatus: Lock Step consciousness stage one: The Tray.

The proposition that accompanies the apparatus is that the microscopic hairs represent some continuous region of the brain (in real life made up of neurons, glial cells and so on—brain substrate if you will) and that by agitating the hairs we can change the shape and position of the junction between the two coloured oils.

The red oil covers an area of my imaginary brain substrate where incoming signals are processed. The far left side of the tray, the red end, represents sensory input. As you travel away from that end, towards the centre of

5. This See, Is the Conscious Bit

the tray, the sensory input from the edge is processed and patterns are picked out. For example, for a visual signal coming from the retina of the eye, the signal might be processed to identify vertical and horizontal lines, and various shapes and, as you travel further and further from the left hand edge, complex recognisable shapes such as a robin or a cat (or whatever animal might be in your field of view) will get picked out from the signal (Figure 8).

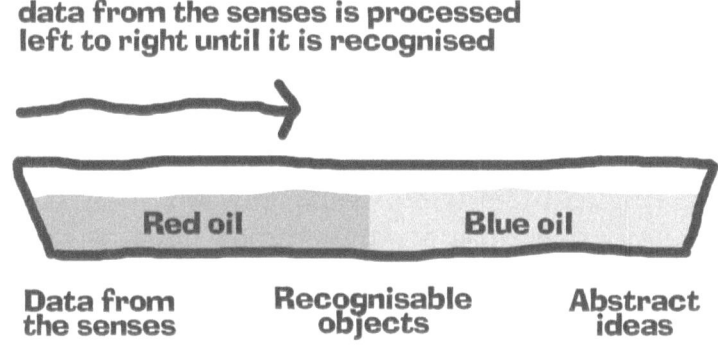

Figure 8: Oil Tray Thought Experiment - stage two

The blue oil covers an area of brain substrate doing the opposite. Under the blue oil the brain has generated a best guess as to what the red oil represents the world as being like. At the blue edge (far right) we have very general, perhaps vague, very abstract concepts. The farther from the blue edge you travel, the more specific the guess, through general animal shapes, to specific instances of robins and cats, and then down to detail of this or that specific feature: a red breast, the extended claws of a paw, a patch of off-white fur.

At the junction between the two oils, the red-side estimate from sensory input meets the blue-side proposal which has emerged from generalities and abstractions. At the junction, the estimate and proposal may correspond to a greater or lesser extent, and the junction itself will move until a satisfactory correspondence is achieved, where the best candidates from each side most closely match each other.

A robin is the best fit or the junction shifts, and settles on something unknown but likely a small russet-coloured bird.

In the Lock Step model the junction is where blue proposed data and red sensorily derived data actually meet, and this is where consciousness resides: the seat of consciousness.

If the red oil covers a substrate that is processing what is probably a robin, the substrate under the blue will be busy trying to find a robin-shape to match the clues and complexity available at the junction between the oils.

So long as there is no match, the fine hairs of the imaginary brain substrate shift the oil junction to and fro, until the red and blue sides arrive at the best match they can of what the content seems to be; the red-side evidence suggests mostly robin-shape; the blue-side conjecture presents a conceptual robin which closely fits the evidence; across the junction (at least locally) the red and the blue adjust to the best guess of a robin: you perceive[1] a robin.

Recall if you will our artist friend, Fred Oblivious. As he works, studying his subject, he shifts attention between more and less detail in his visual impression of a horse. At one moment he focuses on the exact colour of brown in a patch on the muzzle and shortly after attempts to identify the curve of the neck between jaw and shoulder, only to zoom out to compare the length of the neck to the length of the back.

In terms of our thought experiment, the micro-hairs of the substrate in Fred's oil-tank brain are shifting the

---

[1] The model elaborates how consciousness comes to have the content it does; the model does not elaborate how consciousness gives rise to feeling, or what-it's-like-to-be a thinking thing, per se.

5. This See, Is the Conscious Bit

interface between red and blue furiously in search of a match—activity which is most ferocious where Fred's concentration—his fullest attention—is focused.

This thought experiment offers some other interesting features. We find we have ways of thinking about:
(a) Mistakes and optical (or other) illusions,
(b) Frustrating, mind-straining searches for hard-to-find and tip-of-the-tongue memories.
(c) Memories are almost always (re-)constructed—rarely will they be pure, 'photographic'.
(d) The idea that we construct our reality from a paucity of data is clearly illustrated.
(e) We can conceive of a way that attention may be divided when e.g. two or more regions of the junction display heightened activity.

I will call the phenomenon presented in this thought experiment Lock-Step Consciousness, since the idea is that the seat of consciousness is the junction where (an individual brain's best theory of) the evidence meets (an individual brain's best theory of) the abstracted world content. Where brain evidence meets brain speculation. Where the two lock step.

We can further speculate that where attention wanders, i.e. the junction changes shape and moves, under the power of the hairy substrate, the greatest activity in the substrate indicates the highest focus of attention, and that this will move—it will wander—under some kind of perceptual momentum that, I suggest, follows meaning. Attention is drawn to the thing that has most meaning, which in real brain terms (I conjecture) correlates with the highest level of neuron-to-neuron traffic. I'm guessing the junction can only remain stationary—not move at all—in close-to-meditative states.

There are problems for the thought experiment of

course. It is a metaphor. It is not an accurate model, nor any kind of simulation.

(a) In a real brain the oil-to-oil junction would be three-dimensional, for a start.

(b) What reason is there to think the junction is continuous?

(c) While the red edge is sensory input; what is at the blue edge (Pre-conceived ideas, learnt ideas, general abstractions? If so, where do they come from? How are they formed?)

(d) Why should we think the two processes, evidence processing and concept generation are not coextensive[2] in the physical brain? And what of my ribbon of lock-step consciousness then?

(e) Where does causation (and personal agency) lie?

Let me briefly consider another difficulty: How are we to manage time in our Lock Step model? I am not thinking here of the passage of minutes, hours and days, but of the instant of time. We live in the moment. And yet that moment cannot be an infinitesimally small interval. If our consciousness is made up only of infinitesimally small intervals, how do we experience any connectedness, one moment to the next?

Some essential aspect of consciousness must bridge these small intervals, as if our perceptions of things are smeared into one another.

While, at this point in this text, I have no specific answer to that, I think possibly the model can accommodate some sense of sensory smearing.

In the model the junction where red and blue (evidence and speculation) meet is presented as having the shape of a ribbon, 1 cm or so deep. An assumption so far made is that the ribbon junction between the two oils is vertical. But it could be sloped, a meeting of two wedge shapes.

---

[2] Occam's Razor would balk at any duplication of processing

5. This See, Is the Conscious Bit

And what we count as conscious awareness could arise across the sloped side of the wedge, lagging or trailing as the junction moves.

I think also a particular refinement of the model is called for and will be useful.

A simple rectangular tray may represent some small section of the brain, but the brain as a whole might be better represented with our red and blue oils spread across the surface of a sphere. Red oil ebbs and flows from the north pole, where a ring of mountains represents the ring of the body's senses. Blue oil ebbs and flows from the south pole, where only vagueness exists until, moving northward, we see the oil encounters the fine-haired substrate, which is well-groomed and the well-trodden guide of meaningful, if abstract, ideas.

And of course we now have a mechanism for learning. In the infant, blue oil extends from the south pole almost all the way to the north. As the infant is exposed to the world, the red oil pushes out across the northernmost fine-haired plains, shaping them with the patterns that most often occur in the sensory data. The blue oil returns slowly southward, yielding its virgin and unshaped territory, reinforcing and changing the patterns along the junction as it travels, as meanings are realised, and local hot spots of activity are visited again and again and leave preferred tracks in the fine hairs of the substrate to reflect frequent use of common or important patterns.

I will not, for now, elaborate further on the Lock Step Consciousness model/thought experiment which I will call Lock Step consciousness stage two: The Sphere. But I want to have introduced it to aid what follows. However, it is just a model, an incomplete approximation, an abstraction of numerous features, an extended metaphor: a way of trying to capture in one idea the arena for the problem that we are trying to solve: What is consciousness?

Now I turn back to the window.

The robin has gone and the cat is on the snow-covered lawn, staring at me, as if I were to blame.

6. Qualia, the Possible and the Particular

Our sense organs send messages to our brains which in turn generate a range of sense-based feelings, particular to the specialisation of the sense organ, and these feelings contribute to, and are constitutive of, our conscious selves. Feelings such as green and yellow, sweet and sour, lavender and rose and wild garlic—to name but a few—but all of which supposedly represent something beyond the confines of our consciousness.

We are also capable of internal thoughts in the form of unspoken words, and images, and ideas—in our cogitations, and from memory. And more, because we can feel emotions such as love, hate, fear, exhilaration and the rest. And pain.

What prospect is there of explaining this great variety and range of felt things in terms of the Lock Step model? In terms of each small unit of feeling which we have labelled qualia[1], how do qualia come to exist bearing all

---
  1   There is a question here about how qualia constitute consciousness if, as I have implied, consciousness consists of qualia; it is a collection—suitably connected and apparently contiguous—of qualia. If I experience pain, i.e. a qualia of pain, how am I also able to report that pain since, in one sense pain is an extension of the thinking process which reports it. How can qualia be self-refering? Think of it like this: If a candy-sized piece of reflective tinfoil is floating on the surface of a lake, how can any other part of the lake—except that immediately adjacent the tinfoil—"know of" (or recognise or respond or react to—or report) the presence of the tinfoil? If we allow that one portion of the qualate surface can report on another portion, what is the mechanism of at best signalling <pain + apparent physical location> or, at worst, and question-beggingly, 'observation'? Maybe, just maybe, we are going to allow the junction in the Lock-Step model to fold around on itself. To form a hoop, to allow tangents and intersections and even planes to part-way co-extend...

But do we risk stretching the metaphor of the model beyond any functionality it can reasonably give an account of? In the real world, can there

these different flavours?

First, let me invent and give a name to the feature of the universe (or at least of our model) which gives rise to qualia: let me call them gleeons[2]. A gleeon is a fundamental particle in the same way that a quark or electron or photon is fundamental.

The gleeon described here is our first stab at what such a particle might be and do, so this is a gleeon Mark I.

Let me also, immediately, dismiss a vagueness that casts its shadow over the concept of qualia. That is: How big are they? How big is an instance of the redness of the evening sky? Or the squeak of a mouse? In what I propose for the Lock Step model, qualia may be any size you care to think of (literally) but however big or small qualia are for you, the number of gleeons that give rise to an individual quale—like the number of atoms in a wedge of cheese—is astronomical, and whether we are dealing with 2 times ten-to-the-power-twenty-three or 6 times ten-to-the-power-twenty-three is neither here nor there.

So, the brain delivers qualia of many different kinds to contribute to the mind—all sorts of sights, sounds,

---

exist a direct correspondence between the Lock-Step junction and regions of the brain which are 'lit up'; the Lock-Step junction is obviously three dimensional, and could 'come at/towards/reach out to' the pain region from an abstraction such as language, how close does the 'ribbon of pain' have to be to the 'ribbon of language' to be reportable (if not adjacent)? We may be invited, in the real brain, to introduce parallel, co-extensive paths in the brain for things like pain—indeed the blue and red oils might in the real brain need to overlap completely in order to avoid duplication of process—in which case can brain cells operate in two modes: input processing and predictive? If two modes, why not three, or four... How much structure must be imported to serve all the complexities of qualia interaction that we experience? Have we thrown the baby out with the bathwater, merely replacing a model of the brain with a network of statements of how thought, feeling and emotion are connected? If so, this must be wrong. The answer must be simple; Nature demands simplicity. Occam's razor is a Natural tendency, since the most efficient methods in time and energy will out-compete, evolutionarywise, less efficient, more languorous processes. We need a principle, not a list of rules.

2   Glee being a strong and motivating experience.

6. Qualia, the Possible and the Particular

smells, cogitations etc[3]. Are we to understand that each kind of qualia arises from a separate kind of gleeon? Or are all qualia served by a single universal gleeon, but any of those gleeons may exist in one of a variety of different modes or states? Or, does the gleeon simply respond to some feature of its adjacent brain material/substrate (the myriad of fine hairs in the Lock Step model), somehow generating differently flavoured qualia as a result of the interaction between gleeon and some local, specialist property of the brain?

Before exploring the matter of flavours, it does not seem unreasonable to suppose that there are qualia felt by non-human species which must feel wholly different from any of those we might experience. Some animals see light in the infra red spectrum, and some in the ultra violet—what can that be like? Similarly for sounds outside our range of hearing. And what of sharks who can sense electric fields in order to home in on their prey? What can that be like?—surely that feeling cannot borrow from the repertoire of feelings which we humans enjoy since the shark has all the senses we have (sight, hearing, taste, touch and so on) which we might reasonably assume give rise to qualia which are qualitatively similar to our own[4], plus they have this extraordinary specialisation.

The point is, in exploring ideas about how gleeons might give rise to qualia it is important not to limit the possible qualia we give an account of to a catalogue of our own experiences. We should expect there to be many

---

3 Some philosophers have sought to categorise these sensory experiences into primary and secondary qualities. I don't propose to do so at this stage. I do not think the categorisation is needed. I think the categories will emerge automatically from the general model which I present later.
4 This may be a wildly optimistic, or extreme and unjustified, assumption. However the point I wish to make is that I do not wish to impose an unwarranted restriction on the range and type of qualia possible in the universe. My example is intended to dramatically illustrate the point.

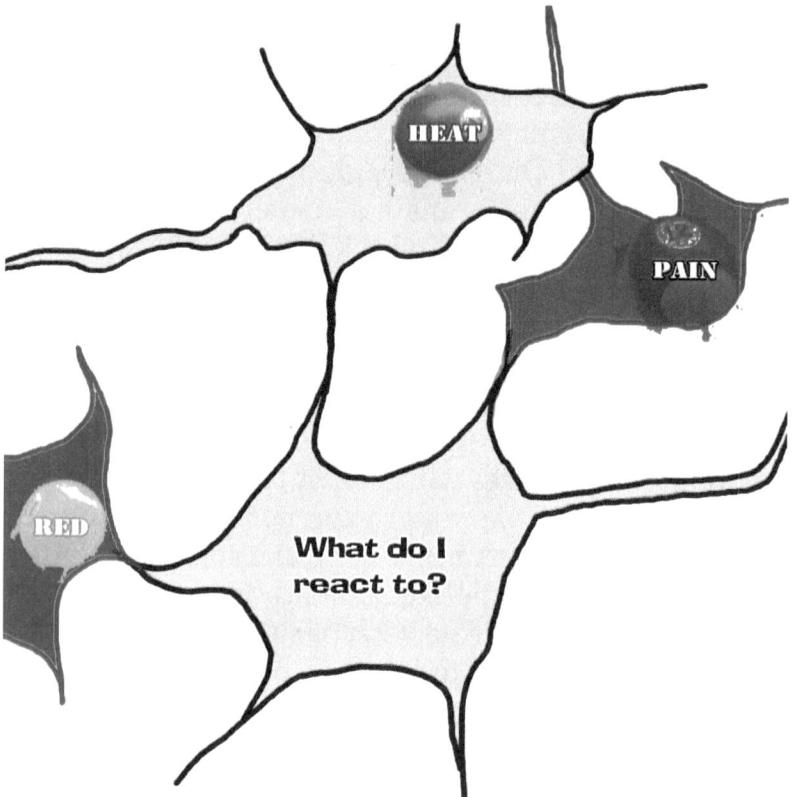

Figure 9: Communicating neurons. Feelings compete to be noticed. But do the neurons that decide what to do respond to other neurons or to the awareness generated by other neurons?

other, likely qualitatively different, kinds of qualia out there in the universe (if only we had a way of capturing them!)

In many ways the gleeon Mark I is no more than the smallest possible homunculus[5]. And thus still begs the question. Without further elaboration I merely invite you to imagine that these gleeons ride our brain cells and react to the content. With this picture in mind, we can ask:

---

5   It is not actually a homunculus because it converts miniature local brain states to miniature local qualia that contribute directly to, and are constitutive of, mental life, but it is like the homunculus in that the question of how does consciousness come about is still begged.

6. Qualia, the Possible and the Particular

(i) What binds all the gleeons so that the experience of the whole seems unified? and

(ii) How do gleeons deliver a broad range of different experiences?

The universe—Nature if you will—is prone to find and use the simplest possible mechanism in its workings. This is why Occam's Razor is an effective tool. And why mathematical models which merely add variables to account for inconvenient data are usually incorrect (They are essentially descriptive and prone to sideline any underlying principle which otherwise, had it been properly honed, would have accounted for all observations on the basis of its essence).

We have already alluded to several possibilities of how gleeons might work. There might be a set of basic gleeon types, each type delivering a different experience. There might be just one type of gleeon but which is capable of operating in different modes. Or the gleeons might be identical all the time, in all respects, but the range and variety of felt content is delivered by local configurations they might adopt (although this last one sounds dangerously like question-begging[6]).

Let us simply assert one of these, suggest a mechanism, and see what insights we may glean (remember this is a model, a *gedankenexperiment*; so I may invent, so long as I do not lose sight of the fact that this is pure invention, and I keep to the rules I create).

Suppose:

(i) all gleeons are identical and that any one gleeon might instantiate any possible feeling that the universe might deliver.

(ii) all gleeons (like the quantum fluctuations in empty space) pop into and out of existence, all the time,

---

6   Because a configuration has likely to be interpreted by someone or something, which is going to be a homunculus.

throughout space, but the duration of their existence is prolonged if they are captured by the right kind of field[7] (I don't know what this field might be, but let us suppose it can be generated by an interaction either between neurons, or within neurons, or a bit of both).

Space is popping with unconnected discrete gleeons[8]. When a gleeon pops into existence inside a neuron at the Lock Step junction (where neural prediction meets processed sense data) it is captured by a suitable brain process[9] and it persists at that location in the brain so long as the Lock Step junction persists[10] at that location in the brain; when the 'ribbon' of the junction moves on, as attention wanders, the gleeons that are trapped at the junction are released (or allowed to collapse), and any awareness taking its content from the location that has just been vacated, evaporates. Awareness moves on, in concert with the junction, borne by the gleeons which are contained by the junction[11].

---

7 The idea here is that I want to avoid bothering with the energy of creation of the gleeon. Why has no such energy already been discovered? Possibly because no one knows where to look for it. Possibly because there is none. Let's see where this latter assumption gets us...
8 In fact it needs to be popping with gleeon, anti-gleeon pairs because there can be no net gain of gleeons out of nothing. What happens to the anti-gleeons, I conjecture, is that while they may have the same property of awareness as their gleeon counterpart, if they are not gathered by neuron activity (except by anti-neurons?), they simply never contribute to any larger conscious awareness (later I suggest that anti-gleeons are pre-conscious, so introduce no conscious component at all).
9 If not spontaneously generated by the process itself. But I seek a pure, simple and extreme scenario here, to make a strong clear point to dance my arguments around. Also, I am deliberately choosing the least complicated, the laziest, if you will, mechanism, in honour of Occam.
10 This could be a fleeting moment, a millionth of a nanosecond, or could last for seconds, even minutes. I don't wish to constrain this time value.
11 The gleeons could be lost or released as the junction moves, instead of being dragged along with it. We need only have some kind of overlap to maintain the illusion of a continuous whole—in a similar manner to the role of memory in persuading us that we are exactly the same conscious person today that we were yesterday.

6. Qualia, the Possible and the Particular

Not forgetting that these gleeons are in many ways the smallest possible homunculi (our question-begging friends), there is then the question of how gleeons deliver specific types of feeling, light, colour, pain and so on.

The answer I propose for Mark I gleeons is, let me warn you, bold.

I draw an analogy from quantum physics: but whereas quantum physics deals in probability, i.e. at the quantum level we cannot know the result of any measurement in advance; all we can know in advance is the probability of observing thus-and-such result. This first version of gleeon theory by contrast deals with possibility. A Mark I gleeon is capable of giving rise to anything that it is possible to feel in the universe and stays in that state of widest possibility until it is forced to settle on one particular feeling (manifesting a microscopic bit of qualia) by some feature of its local context. While its content (what kind of feeling) will likely be determined by some feature of a local neuron, the mechanism by which it is forced to settle its state might be driven by that neuron[12] or by the proximity (intrusion or collision) of a neighbouring gleeon.

I can offer no experimental evidence of this model. I am not trying to. Indeed the model is only a sketch, with suggested alternative details (you have been reading my workings-out as I go along). For the time being, I am merely trying to conceive of an approach which satisfies the criteria I have set out in this and preceding sections, and which could plausibly provide a mechanism for consciousness that does not rely on the fully-fledged machinery of a *Numskull*®-scale homunculus. To my mind at this juncture it is desirable to borrow the principle of fundamental particles from our wider understanding of the universe. Why should gleeons, if such things exist,

---

12   Other flavour-determining mechanisms might be at work e.g. involving context for the neuron. I will touch on those later.

not be on a similar scale as other fundamental 'particles'? Furthermore, in adopting an outrageously generous mechanism for the Mark I gleeon (some might say it is an outright cheat) I am creating a kind of conceptual variable, or placeholder, while I make sure I have the rest of the picture in place. Nonetheless there is something attractive about the idea that gleeons resolve to some particular feeling or other from a range of feelings simply on the basis that in this universe Any Feeling Is Possible and their number and kind are too many to quantify (and possibly infinite).

Let me make two observations about the model.
(i) The junction that moves in response to micro hairs of the substrate in the tray in a Mark I gleeon brain would move in response to, and follow, regions of dense activity.

Where brain cells are most active, there gleeons will thrive and, in this model, I would suggest the highest gleeon density indicates the most focused attention. It makes intuitive sense to suppose that attention at least correlates with high levels of brain cell activity. In this model, activity can shift, as focus can shift, and indeed the direction in which focus shifts will correlate with what appears to be of interest to the individual—I would go further and suggest that neuronal activity and gleeon activity follows meaning, or seeks maximum meaning— the maximum activity in the maximum number of neurons. The focus of consciousness is continually in motion, chasing meaning. This, if true, would yield a theory of meaning, and a theory of attention by side-effect. But of course I am talking only of the Lock Step *gedankenexperiment* and the Mark I gleeon. And I speculate.

Whatever else at some point I might need, I will need my final model to touch the ground—to offer testable conditions under which it might be shown to be false.

6. Qualia, the Possible and the Particular

Falsifiability[13] being a requirement for any theory that seeks the epithet Scientific and seeks to draw upon the benefits and credibility of that epithet.

Clearly I am still some way off that desiderata.

(ii) There is another well-known problem of philosophy which this model offers to shed light on. It is the question of self-reference.

When you report a pain, how is it that the words you speak and the pain you feel are both part of the same consciousness yet the two are distinct? Standardly: how can your mental life comment on itself without becoming a separate observer of itself? (which risks leading to an infinite series of homunculi).

An example given earlier (first footnote of this section) was of the idea that one's mental life is the surface of the ocean, and a particular pain is a small patch of tin foil floating on the surface. How can any part of the ocean that is not adjacent to the tin foil report on the presence of the tin foil? You cannot hoist yourself to the top of a mizzen mast and take a look because as soon as you leave the plain of the sea you become a separate homunculus and immediately beg the question.

The Lock Step model might cope with the pain scenario by allowing the ribbon junction to form swirls, i.e. to come at the pain-bearing sensory data more than once and from different angles. But how can that be translated to real world brain cells?

Movement across the hairy substrate of the Lock Step model is an abstraction. The hairs of the substrate do not correspond one-for-one to neurons or anything like neurons. The purpose of the model is to provide a way of thinking about consciousness and its features not model brain cell activity.

However, the Lock Step model can shine light on the error we make in constructing the problem of self-

---

[13] in tune with the work of philosopher Karl Popper

reference. As follows:

The Lock Step model is a three-dimensional metaphor, and the junction where consciousness is said to emerge is a two-dimensional ribbon. Similarly, the surface of the ocean, with its inaccessible piece of tin foil, imagines a 2-D surface.

These things seem ok, and unproblematic when we model the brain. The brain is 3-D and we can easily extrapolate these models (these ways of thinking) to 3-D, and the problem of self-reference remains. We cannot take a step outside ourselves to acquire another homunculus to do the job.

Indeed the sentences we use to describe these problems are 2-D to look at on the page and 1-D to listen to. This reinforces the impression that a stream of words, e.g. "I have a pain in the bony part of my right ankle." cannot in a single conscious plane both describe the pain and feel the pain; how do the words of the sentence have access to the pain? We find ourselves unable to give an account of the connection.

The error in the argument that you cannot know about the painful tinfoil unless you are adjacent to it (and sentence formation and associated cogitations are not) is that the way that neurons are connected to one another is not 3D (nor 2D nor 1D); neuronal connectivity is massively multi-dimensional.

Consider a cluster of ten thousand neurons. Consider that every neuron in our cluster is connected to every other neuron. Although they exist in three dimensions, they are actually wired in multiple dimensions. If a group of 100 of these connected neurons is dedicated to processing language, and another 100 is dedicated to processing pain and pain locations, it is also true that every language neuron has access to—is adjacent to— every pain-processing neuron. The tin foil on the ocean problem immediately goes away, an artefact of our over-

6. Qualia, the Possible and the Particular

One-dimensional set of connected nodes (a single number can be used to identify any one node)
Each node is adjacent to only two other nodes**.

Two-dimensional set of connected nodes (two numbers are needed to identify any one node)
For our purposes, each node must be adjacent to at least four other nodes (2 for each dimension)**.

Three-dimensional set of connected nodes (three numbers are needed to identify each node)
For our purposes, each node must be adjacent to at least six other nodes**.

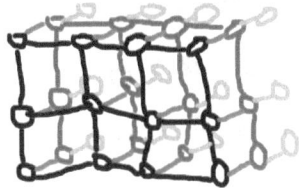

...and the neurons in our brains are massive in number and massively multi-connected, so activity in one 'processing plane'*** may easily connect either massively or in-passing with an entirely different 'processing plane'.

** except those terminating a chain
*** suggested sites for performing e.g. convolutions

Figure 10: Spoken sentences are 1-D (a single stream of data). Read sentences are 2-D (the characters making up the sentence are images). The brain is 3-D (it occupies a volume of space). But brain cells might be connected in one of many dimensionalities. Most likely the multi-dimensionality of the brain, having developed from real encounters with a muddled world and being optimised by nature (which cares not a jot for the convenience of our idea of dimension), will show no discernable dimension-like pattern, but of course the brain works!

simplified, dimension-limited modelling.

In this section we have helped ourself to the Mark I gleeon as a conceptual placeholder and have looked at some of the ramifications of the Lock Step model.

In what follows I will explore further demands we might make of the Lock Step model before returning to the nature of the gleeon, and introducing Mark II.

7. Evolution and Free Will

Is any purpose served by our being conscious?

I think this question can be explored by examining two lesser questions:
(i) Is consciousness needed for brains like ours to work?
(ii) Does consciousness offer evolutionary advantage?

The Lock Step model (of gleeons Mark I) is silent on whether gleeons play a causal role in the mechanism of the brain. The neuronal activity—the fine movable hairs of the model—might continue to operate, to shift the junction between the two oils this way and that, without any reference to any particles that arise from the junction or are trapped between the two oils. So, for the moment, I will put any arguments for or against the necessity of having consciousness for the brain to work to one side.

The question of evolution, however, is more promising.

We have to acknowledge that sometimes evolution makes (apparently random) alterations to this or that species which serve no useful purpose, i.e. alterations for which the most that can be said is that the alteration does no harm, and is redundant in terms of survival. Consciousness could fall into this category. Were consciousness to increase the likelihood of any harms at all, we might expect it to be weeded out. This may be a weak argument in favour of our having evolved consciousness (and unprovable, being a conjectural explanation of an evolutionary trend) but: suppose having consciousness does confer evolutionary advantages, what advantages might those be, and what do they tell us about the nature of consciousness?

I suggest that candidates for evolutionary advantage fall into two main categories:

(i) qualia of pleasure and pain might offer motivation to behave in a certain way, to reproduce more frequently, or to avoid physical harms, or premature death.

(ii) the raw processing of sense data may be eased, or made more efficient by, for example, being able to 'see at a single glance' where some feature of the world offers a threat or an opportunity which we might not be aware of if our 'view' is limited to a localised region of sensory information.

Dismissing the latter first: Given the multi-dimensional nature of the connectivity between brain cells, I do not think consciousness could add anything to the processing already done.

(In computational terms any maxima or minima in, for instance, the visual field, like a bright spot of light, will be located by the wandering cluster at the Lock Step junction, (i.e. as focus of attention shifts), just as well as it might be located by the concentration of neural activity following 'maximum cellular activity'. The Lock Step junction is tied to neural activity and so cannot 'out-perform' neural activity. In the vernacular—if you are distracted, you will miss the obvious anyway. Consciousness does not reliably confer the advantage of 'an overview'.)

However, the question of incentives via strong feelings is quite something else because, surely, pleasure and pain have a piquancy to them that is not available in mere data being processed. And that piquancy will surely feed into and enhance any attempts to achieve any, otherwise base, mechanical goals. Instances of strong feeling motivate behaviour that will deliver either more or less of that

7. Evolution and Free Will

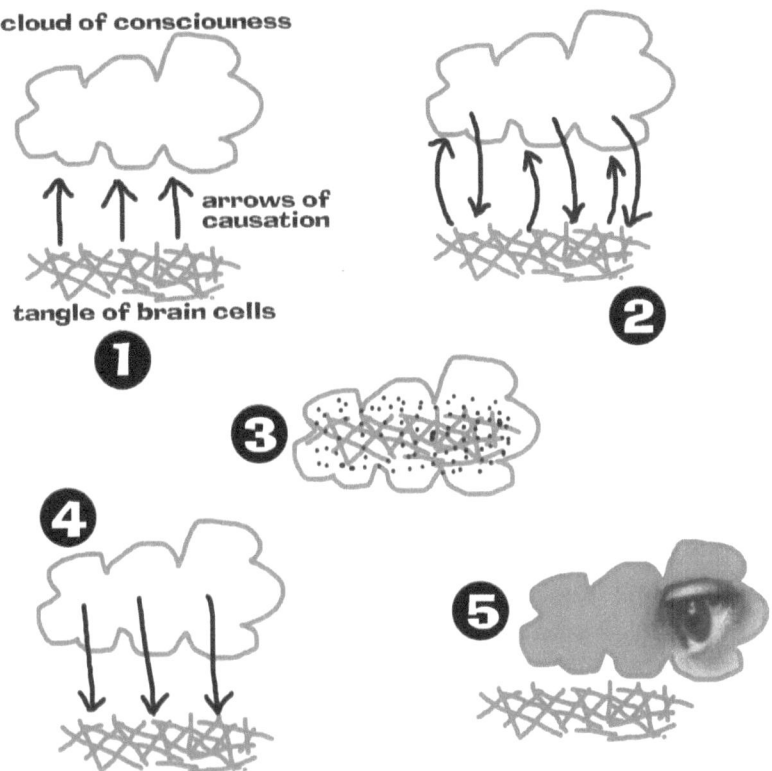

Figure 11: 1-Consciousness is generated but passive; 2-Consc. is participatory; 3-Consc. is identical to brain but under a different description; 4-Consc. dictates to the brain; 5-Consc. is the curious observer.

particular feeling. In evolutionary terms, with regard to survival of the fittest it is entirely plausible that a more highly motivated creature will survive where a less highly motivated creature will succumb to extinction (or at least become a minority in any population).

If we accept this line of argument, and that consciousness delivers evolutionary advantage, we are forced to conclude that consciousness plays a causal role in our behaviour.

This is a significant result and ought to be reflected in our model.

However, there is an interesting, experimentally

observed, twist to this conclusion.

In experiments, human beings are observed as reporting conscious awareness of events up to half a second after the brain is observed to respond electro-chemically to those events. Leaving to one side that the apparent lack of immediacy is odd, there is an implication for the role of consciousness in evolution. The evolutionary advantage conferred by consciousness cannot be related to the speed of a response (to pleasure or pain); speed of response seems to revert to purely neural reflexes. Any evolutionary advantage must be related to processes that take longer to complete than the delay between electro-chemical responses and reported consciousness (indeed has the delay evolved because having a delay itself, and thereby being forced to rely on reflexes, has evolutionary advantage?) Regardless, this suggests that consciousness contributes its advantage in the long term, not the very short term, which means that its advantage must be delivered slowly, e.g. via memory or ongoing cogitation following the experience.

This also has an impact for morality and justice; can we be responsible for what seem like, albeit complex, reflex responses, which we perform less than half a second from any stimulus, or provocation?

Consequently it seems that consciousness does not make us free except to the extent that we can cogitate on the past and decide and ready ourselves to act in a certain way in the future. We do have the opportunity to train ourselves to have morally good reflexes as it were.

The extent even then, to which consciousness is a free agent depends on the extent to which neural activity drives the focus of attention that is experienced, as opposed to having consciousness in some way independently drive the content selection process.

Gleeon behaviour may be random, and en masse, gleeons may follow something like the laws of

7. Evolution and Free Will 61

thermodynamics (in the Lock Step model, they are so numerous, they are best thought of statistically). With statistical underpinnings, they are not determined but are highly likely to follow neural activity in some particular, most probable, way.

The picture of freedom painted here is: Yes, consciousness plays a role in behaviour. There is causal feedback to the underlying brain cells, but it is delayed. Consciousness is tied to neural activity while at best being prone to small statistical fluctuations that result in our behaviour not always being entirely determined[1].

I suggest that the same argument as applies to the evolution of the human brain applies to all higher order animals. If an animal is observed to exhibit pain behaviour then evolution would ensure that those creatures which are actually capable of suffering pain (i.e. are conscious) would have a survival advantage over those that do not, and so members of any such species would predominantly also have consciousness.

But let me go further than pain as a criterion for consciousness. I suggest that any creature that can make sufficient sense of the world to focus on food, to relocate for shelter, and so on—so long as these behaviours are encoded within a suitably arranged complex of neurons, will also have consciousness. They will be open to description under the Lock Step model.

(Note that, as mentioned elsewhere, artificial neural networks based on computations in elaborate digital computers are mere simulations of neural networks. They are all Turing Machine equivalents. In order for something to be conscious it must have a gleeon capture mechanism. A Turing Machine qua Turing Machine cannot be conscious.)

What is the causal mechanism? Memory. How do

---
1   We arrive at a position where our behaviour is neither pre-determined nor free.

we represent it in the Lock Step model? The simplest implementation would be to ensure that some hairs are stiffer than others, notably where patterns of neural and gleeon activity are most often repeated; thus gleeon Mark I activity hardens the hairs—the hair memory—of the substrate (which is a metaphor anyway, lest we forget).

When a few pages ago we asked:
(i) Is consciousness needed for brains like ours to work?
(ii) Does consciousness offer evolutionary advantage?

Our answer to (ii) which is yes, informs our answer to (i) which is, as a result, also yes.

If we accept that the Mark I gleeon plays a causal role in brain activity, i.e. a brain would respond differently to the world in the absence of gleeons, then either the acquisition of a gleeon, or its continued existence for some period of time in the brain, or its dismissal from the brain, must have a causal impact on the neuronal activity of the brain.

Any such impact must, presumably, involve the expenditure of energy.

Now, given that the universe (Nature) works in the most efficient way (Occam's Razor etc.) I propose a mechanism whereby gleeon capture/generation and gleeon release/disintegration can be energy neutral and yet still have a causal effect on the neurons (or whatever brain cells/substrate supports them).

Hysteresis is a phenomena whereby the measured value of a property of an object depends on the direction in which that value changes. For instance, if you apply physical pressure to a rubber ball it may distort in proportion to the pressure you apply. You can plot distortion against pressure on a graph. But when you reduce the pressure, the shape does not return at the same rate. If you plot distortion against pressure as pressure reduces, you get a different curve on your graph.

7. Evolution and Free Will 63

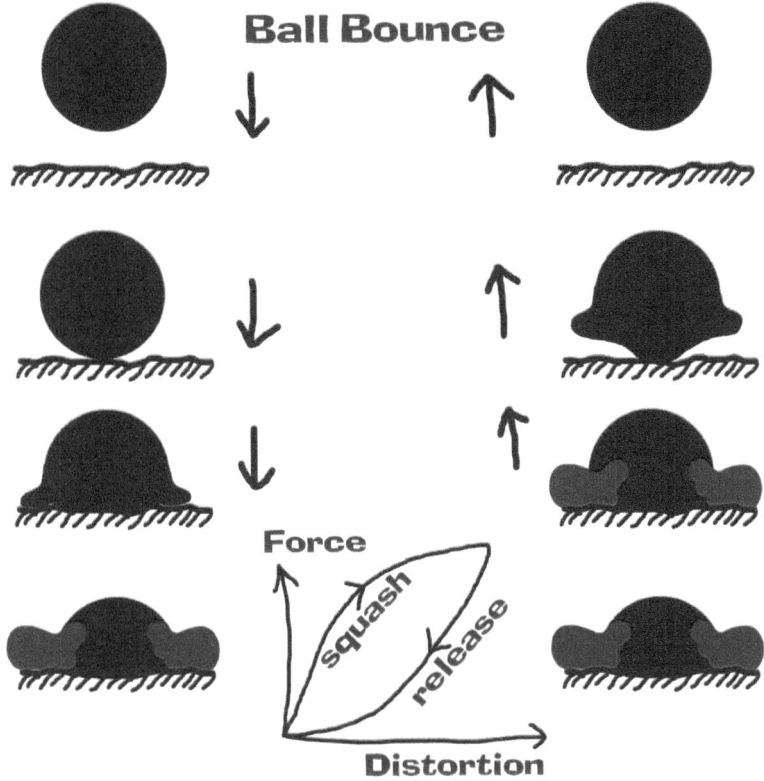

Figure 12: Hysteresis. This rubber ball, after it is compressed, is slow to return to its original shape.

Applying this concept to gleeons, when a gleeon is captured (or created) a gleeonic potential energy is extracted from the brain substrate (i.e. the energy required to create or sustain the gleeon in its captive state). The extraction of energy itself may provoke a memory-inducing change in the brain. But if on dissolution, the gleeon simply returns energy in the same way as it acquired it, the brain substrate may simply revert to its original state and the net effect of gleeon presence is lost. If however the gleeon dissolves (or dissipates or whatever) across a different energy or time profile than its creation (i.e. its demise is not the reverse of its creation), then the brain substrate may be stimulated to respond in a way that it would not otherwise have responded, *sans* gleeon. This at least gives an account of how introducing gleeons into our model—and with them consciousness—could make consciousness a necessary causal component of individual behaviour (not merely some side-effect), which in turn suggests that such a scheme is at least possible. And one might hope plausible. Even if, in the end, the model and accompanying theory proves somewhat wide of the mark.

8. The Good, the Bad and the Choosy

If our model of consciousness is broadly correct; if consciousness is reducible to Mark I gleeons (to what a physicist might ascribe to perturbations in a universal gleeon field) which spring into and out of existence all the time, only sometimes to be captured and sustained by suitable structures in the brain... if this is the case, how can we properly identify our selves with our consciousness because its constituent parts are ever-changing and often (when we sleep) not present at all? Instead, our self is more correctly identified by whatever the continuity of our memory brings with it from the past.

Also, we may not have free will in the sense of being free *tout court*, or even free in our considered actions. However, nor are we predestined to behave one way or another—the generation of gleeons is random, and the behaviour of gleeons *en masse* is subject to the laws of statistics and in that way is probabilistic.

We may not be 'free' in a tabloid sense of "I just decided to do this arbitrary thing for no reason at all." But consciousness gives us autonomy, because it allows us to formulate and 'own' our own rules, based on lived (felt) experience, and allows us to follow those rules.

The fact that we are far, far more likely to behave one way rather than another because of our past experience, via our memories, to a significant degree undermines our culpability for our own conduct. 'We' did not have the leeway to do otherwise. 'We' are 99% determined[1]

---
1   '99% determined' reflects random gleeon generation and statistical distribution of likely, but not determined, gleeon-induced behaviours

Figure 13: With apologies to M. C. Escher: Each of us is shaped, constrained and confined by the society in which we dwell (derived from an original photograph by Swapnil Dwivedi).

and the rest is random. We are to all intents and purposes autonomous—self-governing—but not free.

That said, what would it be like for us, supposing we were free? What would *true* freedom look like?

To be free would be to allow our attention to wander to whatever interests us, or seems important at the time. To make decisions based on our experiences, our memories, our goals and our emotions. To avoid pain—unless the goal demands it. In fact, to be free—to act freely—would be the same and, I submit, *exactly* the same as operating under the Lock Step model.

Does this not mean that what we would do were we free to act of our own volition is precisely what we are determined to do by the Laws of Nature? The behavioural outcomes of free will and pre-determination are in lockstep[2]!

---
2   allowing for random fluctuations at the quantum level and subsequently statistically such that the universe per se is NOT predetermined

8. The Good, the Bad and the Choosy

That may or may not prove a useful outcome for morality.

Society must organise itself as if the individual is responsible and able to choose how to conduct themselves, because society is all about the co-operation of individuals. And co-operation requires a common understanding of (and general compliance with) rules.

Who is the arbiter of morality? Society or the individual—or a bit of both—or must morality appeal to a high ideal that is 'understood' but never fully or adequately articulated?

On what is one to base a morality?

Must not the deliberate torturing to death of an eight year old child be essentially wrong by any measure? Must not the feeding of a starving eight year old child be essentially right?

A moral act (good or bad) may be judged by its effects on the individual at the receiving end, or on the wider impact on society, or by the intentions of the individual doing the deed. Perhaps all of these; there are moral philosophies to match them all (meaning: the question is undecided).

However, what difference does the Lock Step model make?

What counts as ethical conduct towards the Lock Step individual, and what counts as ethical behaviour by the Lock Step individual?

Wherein the seat of the individual who shall be judged? Consciousness, memory, physical body, or an amalgam of the three?

Can a man with no memory be guilty of or rewarded for any actions he has previously taken?

Can a man who is unconscious (or, even, whose consciousness today is not identical with his consciousness yesterday) be guilty of or rewarded for any actions previously taken?

Can a man whose whole life has been reduced to a brain-in-a-vat be guilty of or rewarded for any actions he has previously taken?

I submit that our gut feelings tell us that the culpable individual is the one who has memories that are animated by consciousness.

We value consciousness in ourselves and we extend that value to others who possess it (or, if asleep, continue to have the potential to possess it). A culpable thing is a thing that is capable of change and is conscious.

I introduce 'capable of change', since if the individual was not capable of change up to the point of some heinous act, surely they cannot be culpable. Or, if they cannot change following some heinous act, they cannot be held, thereafter, responsible.

Only a thing that can act freely can properly be brought to book. To punish a creature who cannot change his ways is surely to torture?

This conclusion may not appeal to the revenge-minded[3], but I think it follows from the Lock Step model; the conscious life we value is an active composite.

Turning briefly to the immortal soul and the implication that, possibly, the Lock Step model does away with any need for it (at least for explanatory purposes): Does this not make any and all conscious life more precious and more extraordinary in its uniqueness, and thereby more to be valued?

Here I submit there is an asymmetry: memory per se is deemed to have moral value (as a component of a conscious life) whether or not new memories can be laid down. Whereas those whose memories are immutable (as above, not 'capable of change') cannot be held to account for actions they cannot avoid or change.

Any wider implications than this footnote-sized chapter touches on are beyond the remit of this text since the aim here is to explore consciousness, not morality.

---

3   incarceration without punishment is an option, to protect all

9. Finally, Making It All Work

In the Lock Step model, the Mark I gleeon works like this:
(i) some part of the brain substrate captures a gleeon
(ii) the gleeon connects to some physical property of the substrate which represents content
(iii) the gleeon generates a tiny portion of a quale of feeling which binds to a cloud of adjacent gleeons, at the same time joining its quale to adjacent qualia to create an extended region of consciousness.

We have mentioned but not examined the role of time in the gleeon existence cycle, characterising it only via 'smearing' in order to create overlap while a gleeon is active and to give the impression of continuous and broad awareness.

We have suggested that the brain substrate simply captures stray gleeons (to avoid having to explain the gleeon energy cycle, and simply allowing anti-gleeons to exist unattached, and unconnected i.e. without content and in a pre-conscious state until cancelled out by an encounter with its anti-anti-particle gleeon).

We have not provided an account of what it is that a gleeon has that allows it to generate any part of a quale. Nor have we given an account of how a gleeon is able to give rise to qualitatively different kinds of qualia. To this extent we still have our homunculus, although we have focused in and narrowed down our human-level view to one well-defined difficulty[1] which, if solved, looks

---
1   a good account of what the nascent gleeon does to transition to a micro quale-generator

like it will allow us to build a coherent picture of how a conscious mind might be seated in the brain.

Let me insert here a brief digression on why our conscious selves do not feel like they have any shape in and of themselves. The only shape we can attribute to consciousness is that of the body from whose senses it derives its content. Consciousness has no internal spacial sensors—no space-sensing apparatus inside the conscious envelope of qualia, as it were. The qualia contain no self-referring mechanism. Any apparent self-reference is produced and mediated by brain cells (or whatever are the active components of the brain substrate, most likely neurons). The joint facts that consciousness has no spacial awareness of its self, and that the brain can adopt arbitrarily many data dimensions, means that a 'good brain' should in principle have no limits to the power of its thoughts and ideas. But it will not feel the shape of its own conscious awareness.

The envelope of qualia does not 'think' or 'make decisions' in any positive, active sense. Only by side-effect and indirectly do qualia act, causally, in response to fear, pain, pleasure etc. This action by side-effect may govern patterns of neural activity which have now been adjusted by that prior event of pain or pleasure (This is all that is needed to offer the conscious creature an evolutionary advantage over the unconscious).

However, all that said, it is conceivable that a brain could be so exquisitely attuned to the pleasure, pain etc. that give the gleeons/qualia their causal role, that the decision-making is overt, and affected by conscious states on an observable scale. One wonders though, whether such luxuriating in the realm of consciousness might not equate with neurotic and paranoid behaviour. If this is so, the extent of conscious involvement with brain processing may be self-limiting, i.e. through evolutionary pressure from the other end: too much conscious control renders

9. Finally, Making It All Work

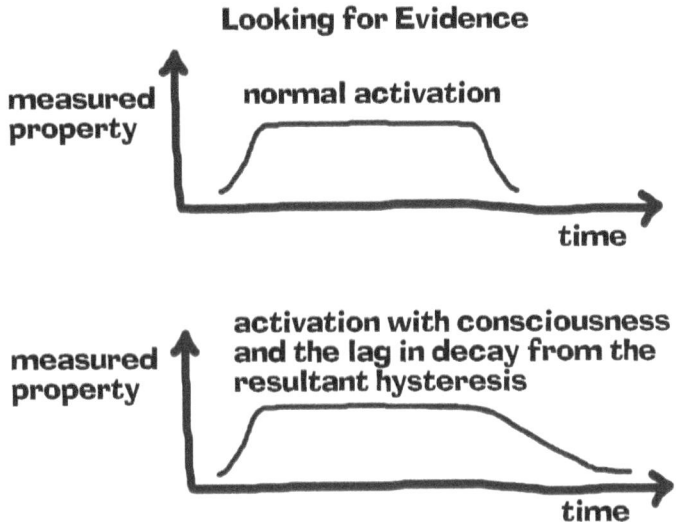

Figure 14: According to gleeon theory, the only physical side-effect of consciousness that science is able to observe is hysteresis in a property that is involved in gleeon capture. This property has yet to be identified; researchers will have to search for it. When brain cells are contributing to consciousness, this measured property will display different rise and decay times than when the brain cells are operating without contributing to consciousness.

the individual vulnerable to unreasonable worry—or an excess of pleasurable distraction—to the point of being vulnerable, evolutionwise.

Let us now see if we can refine the Mark I gleeon and arrive at a more complete Mark II version.

We seem to have arrived at a model where a gleeon is created or captured, and this gleeon is in a pre-qualia-generating state. Once it is captured it is mated with a sensory datum inside the Lock-Step junction. The mated version of the gleeon generates a micro-quale of a kind that reflects the species and variety of datum. The Lock Step junction changes position. The gleeon, being physically bound to e.g. a neuron or part of a neuron, now outside the junction, decays, returning to its pre-conscious state and releases any potential energy it had taken up during the cycle of its capture.

It releases its energy in a way (hysteresis) that can be picked up by the brain substrate and contributes to memory formation (which could not be achieved without the gleeon activation step—i.e. not without consciousness). The period of time for which the gleeon is actively producing its micro-quale is determined by the motion of the Lock Step junction. If there is time-smearing, which our model needs, i.e. to avoid dealing with events of infinitesimally short duration, the gleeon life-cycle must have a finite minimum duration which can be lengthened by a slow-moving, or stationary, Lock Step junction.

I posit that the micro-qualia arising from the gleeons at the junction bind to one another in a similar manner to the coherent light that constitutes a laser (like a waveform across the active field of consciousness), delivering a 'coherent' resonant whole.

I reject the (albeit tempting and ready-to-hand) idea that gleeons are joined by quantum entanglement. Quantum entanglement is a state involving an unresolved system of fundamental particles. However, while they are entangled, one or more observable states of the particles remain undetermined. By contrast a quale is specific, its content is already realised, and that content is being experienced as part of a broader awareness.

Alternatively, if one were to say that it is the collapse of a multi-gleeon wavefunction as its entanglement is broken that creates specific qualia content, then we are back to question-begging: how does the collapse of the gleeon wavefunction give rise to qualia, and how are successive collapses joined to give the impression of a continuous whole (in time as well as across flavours)?

Rather than disappear down that rabbit hole, we already have our mechanism: a free-floating gleeon is captured by a gleeon-capturing field, and bound to a

9. Finally, Making It All Work

datum for some period of time, during which time the gleeon-datum pair generate a micro-quale.

How far down the rabbit hole have we progressed in answering the question *What Is Consciousness?*

Ultimately (and taking the word *quantum* to mean the *smallest possible particle of*) our model and our explanation is one of:

conscious quantum = pre-conscious quantum + pre-content quantum

The conscious quantum is the micro-qualia; the pre-conscious quantum is the gleeon (neither Mark I nor Mark II, but in its final theoretical form); the pre-content quantum is the datum (which remains a conceptual placeholder since we do not know whether it derives from the gleeon capture mechanism, or exists separately, in its own right and is 'collected' from the neuron/substrate).

According to this equation, if we leave consciousness as a mystery, we are positing consciousness as a fundamental property of the universe; we have only succeeded in giving an account of how consciousness might be plausibly incorporated into a sense-data collecting and organising apparatus such as a brain.

Surely we can do better?

Surely we can find a metaphor from everyday life that will convince us of the rightness of the model?

Like what?

Last summer I spent four months living in a caravan in the north of Scotland, close to a small grey stone town called Keith. As a city-dweller, where my horizon is most often concrete or brick and rarely more than fifty metres away, being able to look out of my caravan window across 15 *kilometres* of rolling green hillside under shimmering silver-bright clouds was quite the snaffler of one's life's breath.

Indeed, there was a local high spot (365 metres above sea level) at the top of Meikle Balloch Hill, which I visited numerous times and from which one could see the sea and see the mountain ranges proper, all of twenty kilometres distant.

From that high spot on a good day (between the bouts of rain) the shadows of the clouds shifted across the rolling hills like busy leaves on a lazy stream. This image stays with me and I offer it as the derivation of the final metaphor in this text.

Imagine for me if you will, that rolling green landscape, many kilometres in scale, populated by green fields for grazing and golden barley for harvesting, and there are forests too—every hill of any size is home to a small forest. Now, across this imagined landscape let us invent two great weather fronts, two great sky-covering cloudscapes that are separated by a thin gap of clear fresh air, perhaps only a few hundred metres wide, and through which the rays of the sun pour, and they paint a thin band of bright golden light across the land.

You have this grand landscape picture in mind? Good! Next we play with time.

For suppose, in this dreamed-up world, nothing can live without light. Without light only seeds can survive. As the golden light of the sun shifts across the land, in its wake the trees die, and in its path grass grows, flowers bloom, and trees mature with accelerated vigour—the nature of this strange world has determined that the lifecycle for a plant shall be short. And yet the wild horses, deer, badgers, foxes, birds, and indeed all wildlife follow the sunlight, keeping always within the ever-renewing brightly lit strip of forest.

The band of golden sunlight crosses the land carrying with it all the life of the land, until it comes again to this

9. Finally, Making It All Work

place, as it will, but who knows when?

You get the picture? I hope so. It's the sort of image that is suited to an animated children's movie, I suspect.

This is the home of my end-of-book metaphor. In this metaphor, **life** stands in the place of consciousness; **sunlight** is the gleeon flood; the **seeds of the plants** are the datums of content. It is the Lock Step model brought to life.

Using life as a metaphor, you might think begs the question. But it does not. Life is a mechanical thing. Life is the successful chemical expression of DNA[2]. It is complex and wonderful and capable of infinite variety. That is what consciousness is like. The seeds from the plants laid down in their accelerated lifetimes shape that life, and yet are able to evolve (laying down new 'memories'). There are species of plant (our metaphors for sounds, images, etc.) and varieties within species (our metaphors for flavour, pitch, colour and so on)[3]. The animals are longer lived than the plants, and mobile, and belong to a higher order of things (like abstract ideas). And the photons of light from the sun: well, they are just a gift of Nature, full of potential; the energy of a photon is pre-*anything-you-like*.

Perhaps the metaphor is too complicated. Did you really want a hammer striking a nail?

In what I give you, the whole is greater than the sum of its parts. Isn't that the truth about consciousness? The strangeness of it—the wonder of it—far outstrips any explanation of it one might give.

---

2   Arguably, life includes consciousness, and consciousness contributes (my symbolic guess of) 1% of the causal story. To the extent of 1% I might be begging the question. However, after a few cycles of question-begging my 1% has become 0.0001%, and will effectively vanish into something close to zero as a lasting consideration. I do not so much beg the question as beggar the begging of the question.

3   Primary and secondary qualities as philosophers would name them.

It only remains for me to offer two questions to the scientists that I may show that the theory is capable of falsification and thereby gains the associated credibility.

(i) Can we spot the postulated hysteresis? Maybe this can only be answered when we can give a full account of cellular activity in the brain and in doing so come across measurements or observations that are not explained by what is visible to us under the microscope.

(ii) What makes the difference content-wise between brain cells in the visual cortex, and auditory cortex and other sensory processing areas, such that we have distinct datums to individuate different conscious experience? In particular are there physical features or parameters exclusive to each brain region, or must content come from context in respect of all those neurons (and their internal states) i.e. neural configuration on a macro scale? (If the latter, then the gleeon theory of this text has been undressing the wrong homunculus).

It is only a matter of time of course before the science of this subject-matter catches up with the philosophy of it. One might suppose.

9. Finally, Making It All Work

WHAT NEXT?

I have advanced a theory that fits all the constraints set out in the preceding pages. I have offered two views of the Lock Step model: a particle equation and a grand metaphor.

However, the Lock Step model is silent on how we imagine things that are not before our eyes—or even how we perform actions, like walking.

My next challenge for the Lock Step model is to use it to develop an artificial intelligence that is capable of honesty and morality—and performing actions. This development is a whole new story.

In trying to create an AI based on the Lock Step model and at the same time managing to avoid re-introducing the homunculus, it will be necessary to locate all qualia on a two-dimensional surface (the metaphorical cinema screen of our 'inner eye').

But without a homunculus to interpret what is on this screen, the big question is: How can we perceive the passage of time and the existence of a third dimension when we live only in *The Now* and are confined to 2D?

By solving this problem we can hope to arrive at a machine that is capable of entertaining concepts while never being conscious. A serendipitous side-effect is that it automatically provides us with a mechanism for incorporating honesty and morality.

The title of the book is: **This Robot Brain Gets Life (Making AI Pseudo-Conscious):** *Design Alignment In, Design Hallucination Out.*

What next for the Lock Step model? Let it deal a blow to the nastier problems of Big AI.

C.B. 2023

Acknowledgments

My thanks to Jack Calverley for cover artwork, illustrations and book design.

And thank you to Sanja Baletic and Julian Dixon for their help and support during the creation of this work.

About the Author

Carter Blakelaw BSc BA lives in bustling central London, in a street with two bookshops and an embassy, any of which might provide escape to new pastures, if only for an afternoon.

Blakelaw has studied physics, philosophy and computer science at degree level, was the architect and lead programmer for the Rooms 3D Desktops virtual reality engine, and has worked in integrated circuit design.

www.carterblakelaw.com

Also from **www.thelogicofdreams.com**:

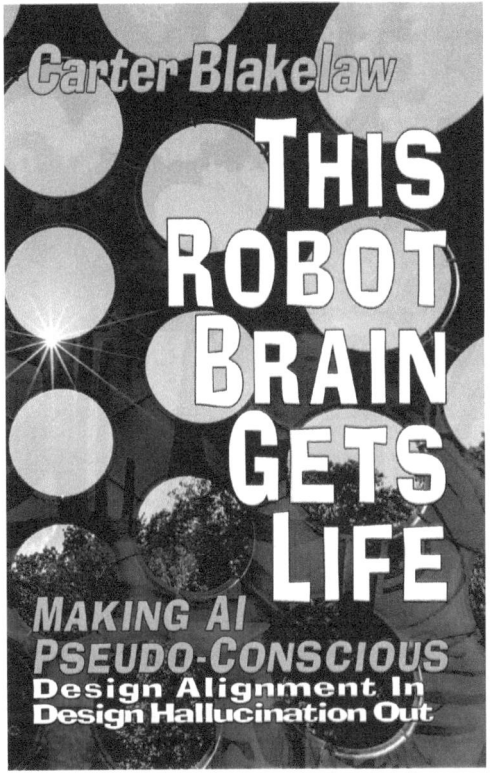

To align an AI's goals with our own, we must build-in alignment from the start.

To keep an AI honest, we must build-in honesty from the start.

To get an AI to understand anything, we must invest it with something of what it's like to be conscious.

In this book, a theory of consciousness is cast into an AI architecture that allows interventions in concept formation *by design*.

Philosophy, psycholgy and computer science come together in a unique take on thinking machines.

Also from **www.thelogicofdreams.com**:

Given a theory of what makes us conscious and, on the back of that, a theory of how far AI can go without being conscious, we can ask what a human being can do that an unconscious AI cannot.

Even then, an AI will likely imitate those all-too-human capabilities too, so is there some realm, some aspect, some art—some corner of the universe that will forever remain the preserve of the human being?

This book examines what is left for the artist of any kind: painter, poet, musician or prose monkey—for monkeys we all are, are we not, in the end?

Turn the page on the challenge for creativity; you want The Authentic? *Then read on…*

Also from **www.thelogicofdreams.com**:

Make Words Tell Tales

A nuts and bolts guide to crafting words into great sentences and great stories. For anyone wanting to hone their craft, drawing 100+ useful rules of thumb from more than a decade of writers' workshops.

You will discover:
    100 rules of thumb to apply to your fiction
    The motivation behind each rule
    The pros and cons of keeping—or breaking—the rules
    Numerous examples of rule-keeping and rule-breaking
    How every rule helps keep the reader reading

If you are serious about your craft, *act now!*

Also from **www.thelogicofdreams.com**:

Four strangers are lured to three houses in Morricone Crescent at the heart of London's Notting Hill where four carefully staged deaths tie them together.

SANDY is delving into an unsolved hit-and-run. He doesn't mean for the witnesses to start killing each other.

LINDA is determined to pay her moral debt to the manager who gave her her first break.

MOE is the journalist who lost his Fleet Street job after asking the wrong questions.

ANGELA, recently conned out of her life savings, all she wants is her money back.

Also from **www.thelogicofdreams.com**:

15 all-new stories

From both new and established, award-winning and best-selling authors
   Kenzie complains about cats to the wrong neighbor
   Mitch chooses the wrong couple to spy on next door
   But does Sheila target the right man to scam?
   Prepare for murder in many guises…

Visit a world where the intolerable few, who create hell for the rest, get their comeuppance.

Short stories of murder, mystery, and revenge from Hilary **Davidson**, Steve **Hockensmith**, L. C. **Tyler**, Marilyn **Todd**, Dave **Zeltserman**, Warren **Moore**, Robert **Lopresti**, Nick **Manzolillo**, Kevin **Quigley**, Eve **Elliot**, Eve **Morton**, Kay **Hanifen**, Wendy **Harrison**, Shiny **Nyquist**, and F. D. **Trenton**.

www.ingramcontent.com/pod-product-compliance
Lightning Source LLC
Chambersburg PA
CBHW021950160426
43209CB00030B/1903/J